Predator Ecology

Predator Ecology

*Evolutionary Ecology of the
Functional Response*

John P. DeLong

*School of Biological Sciences, University of Nebraska-Lincoln
and Cedar Point Biological Station, USA*

OXFORD

UNIVERSITY PRESS

Great Clarendon Street, Oxford, OX2 6DP,
United Kingdom

Oxford University Press is a department of the University of Oxford.
It furthers the University's objective of excellence in research, scholarship,
and education by publishing worldwide. Oxford is a registered trade mark of
Oxford University Press in the UK and in certain other countries

Published in the United States of America by Oxford University Press
198 Madison Avenue, New York, NY 10016, United States of America

British Library Cataloguing in Publication Data

Data available

Library of Congress Control Number: 2021937953

ISBN 978–0–19–289550–9 (hbk.)
ISBN 978–0–19–289551–6 (pbk.)

DOI: 10.1093/oso/9780192895509.001.0001

Printed and bound by
CPI Group (UK) Ltd, Croydon, CR0 4YY

Contents

Prologue

The motivation for this book was three-fold. First, I personally wanted to learn more about functional responses. I found, however, that information about functional responses in the literature is piecemeal. In no place could I find a synthesis about them, despite the existence of thousands of papers describing or parameterizing functional responses for all manner of predators and prey. Second, there was clear conflict in the literature about what models to use to describe functional responses, the biological meaning of the model parameters, and why functional responses vary among predator–prey pairs and across environmental or trait-based gradients.

Third, and perhaps most importantly, the functional response became the core concept for my field-based course called *Predator Ecology*. I wanted to provide my students with an overview of the fundamentals and the biological relevance of functional responses, so that in short order they could interpret papers, conduct their own experiments, and grasp how natural selection might be shaping predator–prey interactions and therefore food webs. I needed to start synthesizing for the course, resolve conflicts in terminology and models, and help students connect the math to the biological reality of nature. That was the birth of this book.

So for me and anyone else, this book covers the fundamentals and then offers a deep dive into what functional responses really are, how to think about them, why they are relevant to pretty much anything ecological, and where studies on functional responses might go in the future. This book is what I needed when I started teaching the Predator Ecology class. This book is intended for advanced undergraduate students and graduate students, as well as anyone interested in functional responses. The book moves between simple introductions, derivations of the core models, reinterpretations and clarifications of the parameters and the functions themselves, and novel hypotheses about functional responses and their consequences. For anyone mostly interested in the concepts and biological relevance, it may be useful to skip over some of the derivations and focus on the biological meaning of functional responses and their parameters. Then come back to the equations later.

To support hands-on learning as well as new research into functional responses, the book is accompanied by a full set of code to reproduce all data and analysis-based figures in the book. This code is written for Matlab

© but could be translated to other scientific programming languages. The code, and associated data as necessary, are hosted in a zipped folder at www.oup.com/companion/DeLongPE. Any corrections or updates to the code will be posted at this site.

This book would not have been possible without the patience and support of all the people who have taken and TA'd my Predator Ecology class both on City Campus at The University of Nebraska-Lincoln and out at Cedar Point Biological Station. Their involvement has helped to keep the momentum on my predator ecology research going. I appreciate the helpful comments on drafts of this manuscript from Stella Uiterwaal, Kyle Coblentz, and Mark Novak. I could not have understood the root sum of squares expression (see Chapter 3) without the help of Van Savage. I'd like to give a shout-out to HawkWatch International (www.hawkwatch.org), where I got my start in predator ecology shortly after college. Finally, none of what I do would make sense without the love and support of Jess, Ben, and Pearl, who make my heart soar like a hawk.

1
Introduction

Predators seem to be universally fascinating. Maybe that is because we humans are predators, or because we can be prey for other predators (Quammen, 2004). Maybe we are just morbidly curious about death. Whatever the reason, I have noticed that nature shows tend to focus a lot on predator–prey interactions: the drama of the predator's hunt or the relief of the prey's escape. We are drawn to predation and intuitively understand that it is a fundamental part of nature. Indeed, what a predator eats is often among the first things that we learn about it, suggesting that who eats whom is among the most central features of ecological systems, or at least central to the way we imagine them (Sih and Christensen, 2001).

Predation is fundamental beyond the event of a predator capturing prey. The rate of predation and the identity of the prey combine to direct the flow of energy through ecological communities. As a result, predation plays a key role in structuring food webs. Of course, there are other ways that energy flows through communities that do not involve predation, such as photosynthesis, herbivory, parasitism, decomposition, and the consumption of detritus or nectar. These are all equally crucial, but consuming other organisms is a widespread way of getting that energy, so understanding the rate at which predators consume prey is a necessary part of understanding ecological systems.

So what controls the rate at which predators consume prey? Many things. Predator traits like claws, prey defenses like camouflage, habitat complexity, hunger, and the presence or behavior of other predators all play their part. Among these many influences, one of the biggest factors is the number of prey available to be consumed. Generally, predators have a higher foraging rate when there are more prey to be had—up to a point. The relationship between foraging rate and prey abundance (or density) is known as the functional response (Holling, 1959; Solomon, 1949) (Figure 1.1).

The functional response is a description of how many prey a predator would be expected to eat given a particular amount of prey available to the predator, wherever they are searching for food. That expected number depends on the behavior and morphology of both predator and prey in the context of

Predator Ecology: Evolutionary Ecology of the Functional Response. John P. DeLong, Oxford University Press.
© John P. DeLong 2021. DOI: 10.1093/oso/9780192895509.003.0001

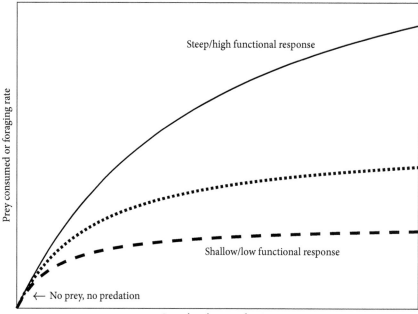

Figure 1.1. Some generalized functional responses. Foraging rates must be zero when prey are absent, so functional responses are anchored at the origin. The curves increase as prey increases, but the shape of that increase, the presence of an asymptote, the overall height, and possible bends in the curve all depend on the specific conditions, morphologies, and behaviors of the predator and prey involved.

a particular habitat, so the functional response is really an emergent property of the total foraging process (Juliano, 2001). The functional response is always anchored at the origin—no prey, no predation (Figure 1.1). There also is always some increase in foraging rate as prey increases. The shape of that increase, and both an explanation for and description of that shape, depend on many factors. These factors relate to how predator and prey move and encounter each other, how predators choose what to attack, how good prey are at escaping, and what other things predators have to do with their time besides hunting.

Functional responses have a long history in the scientific literature (Jeschke et al., 2002). On the empirical side, there are well over 2,000 functional responses measured for hundreds of predator–prey pairs. These functional responses have been digitized and archived in the recent FoRAGE database (**F**unctional **R**esponses from **A**round the **G**lobe in all **E**cosystems) (Uiterwaal et al., 2018). These measured functional responses cover numerous taxa from unicellular foragers of bacteria and phytoplankton (Roberts et al., 2010) to wolves (Jost et al., 2005). A key goal of this literature is understanding what determines the shape of the functional response for any given predator–prey interaction.

On the mathematical side, many models have been proposed to describe functional responses (Akcakaya et al., 1995; Beddington, 1975; Crowley and Martin, 1989; DeAngelis et al., 1975; Denny, 2014; Hassell and Varley, 1969; Holling, 1959; Skalski and Gilliam, 2001; Tyutyunov et al., 2008; Vallina et al.,2014). With these models, functional responses have been used to predict prey selection (Charnov, 1976; Chesson, 1983; Cock, 1978) and invoked as building blocks of community and food web models (Boit et al., 2012; Brose et al., 2006; Guzman and Srivastava, 2019; Petchey et al., 2008; Rall et al., 2008; Rojo and Salazar, 2010). These different forms of functional responses have been argued over and compared for their implications and construction (Abrams, 1994; Abrams and Ginzburg, 2000; Houck and Strauss, 1985; Skalski and Gilliam, 2001). Yet, we still do not have resolution about what mathematical expressions in functional responses are most useful, most biologically meaningful, or most general.

1.1 Functional responses and food webs

Even with the vast amount of work already conducted about functional responses, I would argue that we have only just begun to understand the causes and consequences of quantitative variation in functional responses. This limited understanding is true in general, but it is even truer when it comes to functional responses as they occur in the field. Ecological communities contain numerous species and uncountable numbers of individuals. The number of predator–prey interactions in even a relatively simple food web can run into the hundreds. In a typical cartoon food web (Figure 1.2), just a handful of species leads to numerous pairwise foraging interactions among species. Of course there are far more species in a real food web than shown in the cartoon, and there may be separate functional responses for each type of insect eaten by the shrews or for each type of grass eaten by the caterpillars. We typically draw such feeding interactions with an arrow pointing toward the forager. Thus, herbivores also have functional responses, as do all the interactions leading up to the top predator, in this case a barn owl (*Tyto alba*). For this reason, all of what follows in this book may apply to organisms consuming plants in part or in whole (e.g., algivores) as well as the carnivorous mammals that often come to mind when we think of

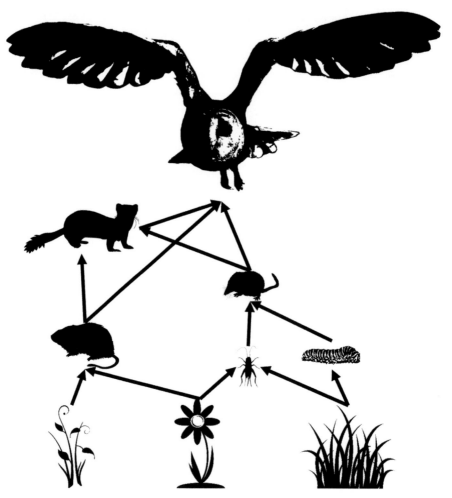

Figure 1.2. A simplified food web. Functional responses characterize the prey-dependent foraging rate for all predators in the food web including the herbivores, not just the top predators. In this way, functional responses set the rate of energy flow in food webs from plants to apex predators.

predators. Predators may even forage on biological entities that cause disease, such as viruses and other parasites (Anderson et al., 1978; Thieltges et al., 2013; Welsh et al., 2020), expanding the relevance of functional response beyond the typical food chains described in ecology textbooks. All of these interactions can be described by functional responses, and chances are they are mostly quantitatively different from each other (Jeschke et al., 2002). They may even be quantitatively different among individual predators within a population (Hartley et al., 2019; Schröder et al., 2016; Siddiqui et al.,2015). Yet, to my knowledge, there are no cases where the functional responses of all the connected species in a food web have been characterized completely, although researchers have done more to estimate strengths of interaction among species in food webs in other ways (Gilbert et al., 2014; Wootton and Emmerson, 2005). Thus, we may have long lists of prey types for many predators in food webs—who eats whom—but we generally do not know the rate of predation on most if any of those prey types. Without this information, we are typically restricted to making taxonomic or body size-based assumptions about the shape and height of functional responses in food web models (Boit et al., 2012; Brose et al., 2006; DeLong et al., 2015; Rojo and Salazar, 2010; Schneider et al., 2012).

Because functional responses determine the flow of energy from lower to higher trophic levels (Figure 1.2), they strongly influence rates of birth and death and consequently the size of predator and prey populations. Variation in the shape of a functional response therefore can have crucial effects on the dynamics of populations. For example, a steep and high functional response (Figure 1.1) means a lot of foraging and a chance that the predator can over-exploit the prey, leading to instability or population cycles (see Chapter 4). Thus, the shape of the functional response is important. Even slight changes to functional responses can lead to big changes in the population dynamics of trophically interacting species (Chapter 4) (Oaten and Murdoch, 1975; Williams and Martinez, 2004). Overall, it is thought that functional responses should be on the shallow side in nature, meaning that no particular predator–prey interaction dominates all of the energy flow (McCann et al., 1998). Such "weak" interactions minimize overconsumption and encourage persistence of predator and prey populations. We will see in Chapter 6, however, that natural selection is not likely to favor shallow functional responses for predators because shallow functional responses limit energy uptake. Understanding what pulls functional responses up or pushes them down is a central problem in the evolutionary ecology of predator–prey interactions and food webs as whole.

Because of their effect on population dynamics, functional responses play a role in virtually all types of community dynamics (Houck and Strauss, 1985; Rosenbaum and Rall, 2018), including trophic cascades (DeLong et al., 2015; Levi and Wilmers, 2012; Ripple and Beschta, 2012; Schneider et al., 2012), predator–prey cycles (Jost and Arditi, 2001; Korpimäki et al., 2004; Krebs et al.,2001; Oli, 2003), and keystone predation (Paine, 1966). More than those dynamics, predation plays a role in determining the magnitude of ecosystem functions that many prey organisms conduct (Curtsdotter et al., 2019; Koltz et al., 2018; Wilmers et al., 2012). Predation is also dependent on temperature (Burnside et al., 2014; Dell et al., 2014), and so the impact of climate change on ecological systems will in part stem from how changes in temperature and precipitation alter the steepness and height of functional responses (Binzer et al., 2012; Boukal et al., 2019; DeLong and Lyon, 2020). Indeed, it is clear that temperature itself is a major driver of the shape of functional responses (Daugaard et al., 2019; DeLong and Lyon, 2020; Englund et al., 2011; Islam et al., 2020; Uiterwaal and DeLong, 2020; Uszko et al., 2017), indicating that functional responses will play a key role in determining the effects of climate change on ecological communities.

The functional response also plays a key role in species conservation and management. For example, introduced predators are generally thought to have steep and high functional responses in their invaded range (Alexander et al., 2014). This high functional response may arise because of a mismatch between the offenses of the novel predator and the defenses of the potentially naïve prey, giving predators the edge in foraging interactions and a chance to survive in their new communities. Without that boost in foraging, out-of-place predators might not be able to persist and invade, since the functional response controls the energy flow to the predator and thus its population size. This hypothesized tendency may be one reason invasive predators, when they do establish, can be particularly destructive to a wide range of prey (Doherty et al., 2016). Similarly, biocontrol predators are released to control the populations of problematic insects, including agricultural pests and infectious disease vectors (Saha et al., 2012; Tenhumberg, 1995). An ideal biocontrol predator is one with a high functional response on the target pest, as a shallow functional response implies little mortality of the pest (Lam et al., 2021; Monagan et al., 2017). Indeed, we can even use estimates of the functional response to show that other pest control strategies such as insecticides can alter the effectiveness of the predator (Butt et al., 2019). Similarly, the functional response influences how well microbial predators and zooplankton can control toxic red tides in marine systems (Jeong et al., 2003; Kim and Jeong, 2004). Finally, humans compete with non-human predators for prey such as game birds. Understanding the functional response can be a crucial part of devising management strategies for native predators that maintain the availability of harvested prey species (Smout et al., 2010). In fisheries, the functional response is needed for understanding and managing both harvested and unharvested fish populations, as the functional response influences the ability of fish populations to grow (Hunsicker et al., 2011).

The generally positive slope of functional responses also suggests that any behaviors a predator can use that can increase prey density can increase foraging rates and thus individual fitness and population growth. For example, northern shrikes (*Lanius borealis*) have been observed singing in winter, with the effect of drawing in potential passerine prey, raising prey density and thus foraging rates (Atkinson, 1997). Small forest falcons also may use calls for luring prey with the same effect (Smith, 1969). In addition, searching behaviors or nomadism would allow predators to increase foraging rates by actively locating areas (local patches) that contain many prey (Korpimäki and Norrdahl, 1991). Thus, movement behavior can allow predators to alleviate the constraint of the functional response, sometimes referred to as a "numerical" response, but this will not be a focus of this book.

In short, the functional response describes an essential process in ecology, with ramifications for nearly every aspect of natural systems and our ability to manage them for ecosystem services, pest control, or conservation. The remainder of this book therefore takes a deep dive into functional responses. Chapter 2 covers the mathematical basis of functional responses, shows the derivation and meaning of the standard models, and standardizes terminology. Chapter 3 describes known variation in functional responses within and across species. Chapter 4 shows how variation in functional responses influences population dynamics. Chapter 5 expands the discussion to multi-species functional responses, which are the functional responses of a predator feeding on more than one prey type. Chapter 6 shows why and how natural selection should act on traits that influence functional response parameters. Chapter 7 describes optimal foraging theory in light of functional responses and how we still do not understand foraging strategies (optimal or otherwise) with respect to fitness. Chapter 8 introduces the idea of prey switching and shows why functional responses are necessary for understanding or even identifying prey preferences. Chapter 9 discusses the many possible origins of sigmoidal functional responses. Chapter 10 discusses the nature of functional response data and reviews statistical concerns in curve fitting. Chapter 11 suggests critical areas of research needed to understand more fully functional responses and their consequences for ecological communities.

2

The Basics and Origin of Functional Response Models

This chapter is the essential beginner's guide to the functional response, its derivation, the various forms, its connection to other models in the literature, and what the parameters mean. It is the ground floor for the rest of the book. Surprisingly, our understanding of the functional response as presented in the literature is quite muddled, with confusion ranging from the terminology used, to the various mathematical forms the functional response takes, to the biological interpretation of functional response model parameters. I provide a summary and forward-looking perspective on these issues.

2.1 Types of functional responses

Functional responses are curves relating a predator's foraging rate on a specific prey to the availability of that prey (Denny, 2014; Holling, 1959, 1965; Juliano, 2001; Solomon, 1949) (Figure 2.1). The foraging rate must be zero when there is nothing to eat, so functional responses are anchored at the origin and have a positive slope at low prey density. As prey density[1] increases to higher density, the functional response usually does one of three things: it may continue to rise linearly (this is called a type I, linear functional response; Figure 2.1A); it may approach an asymptote (type II, saturating functional response; Figure 2.1B); or it may increase very slowly at first, then increase more quickly, and eventually approach an asymptote (type III, sigmoidal functional response; Figure 2.1C). Although it may be that functional responses show more variation than these three types (Abrams, 1982), and numerous alternative formulations have been derived for various purposes (Gentleman et al., 2003; Jeschke et al., 2002; Tyutyunov et al., 2008), this simple classification is still in

---

[1] In this book, predator and prey numbers will be presented as either abundances (just a number in a space) or as density (a number per space). Although sometimes they are presented in units of biomass, it is important to remember that predator and prey are individual organisms, and the impact of predation on population dynamics and evolution occurs through the gain or loss of individuals.

Predator Ecology: Evolutionary Ecology of the Functional Response. John P. DeLong, Oxford University Press.
© John P. DeLong 2021. DOI: 10.1093/oso/9780192895509.003.0002

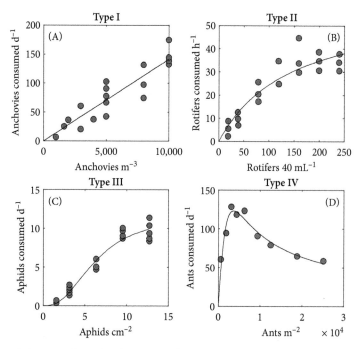

Figure 2.1 (**A**) A type I functional response for the comb jelly (*Mnemiopsis leidyi*) consuming anchovies (*Anchoa mitchilli*), fit to equation (2.1). Data from Monteleone and Duguay (1988). It is possible that this functional response is type I, but it also may be just incompletely determined. (**B**) A type II functional response for the copepod *Mesocyclops pehpeiensis* foraging on the rotifer *Brachionus rubens*, data from Sarma et al. (2013) and fit to equation (2.5). (**C**) A type III functional response for the ladybird beetle *Eriopis connexa* feeding on the aphid *Macrosiphum euphorbiae*, data from Sarmento et al. (2007) and fit to equation (2.7). (**D**) A type IV functional response of the spider (*Zodarion rubidum*) foraging on ants (*Tetramorium caespitum*) and fit to equation (2.9). Data from Líznarová and Pekár (2013).

wide use and covers much of the ground necessary to understand functional responses and what they represent (Real, 1977). Beyond the three main forms, some suggest that there is also a type IV functional response (Figure 2.1D), with the foraging rate dropping again at very high prey density, but this form has not been documented many times (Baek, 2010; Jeschke et al., 2004). Although functional responses also exist for herbivores (Andersen and Saether, 1992), this book will focus on the functional responses of predators, that is, consumers that kill their prey, even if they only eat part of the prey. The main reason for making this choice is that by not killing the plant, an herbivore has a fundamentally different effect on the population dynamics of its plant resources. Functional responses also exist for parasitoids that stun prey and lay eggs in their prey, but these will similarly be less of a focus here.

Typically, we estimate the shape of a functional response experimentally. The usual approach is to measure foraging rates with a series of foraging trials in which an individual predator forages on some arbitrarily assigned number of prey for some prespecified amount of time. The researcher then counts the number of prey remaining and assigns the difference between the prey offered and the prey remaining to predation, trying to account for the possibility that some prey may have died during the experiment for reasons other than predation. Then, the type of functional response can be determined by fitting a model to the data with a variety of methods (see Chapter 10). Here, a model is just an equation describing a process or pattern that we believe to be relevant to our data—we will look at many of these shortly. For example, in Figure 2.1 I fit models describing type I–IV functional responses to each data set, and we can see that, with the right parameters, the equations describe the shape of the data reasonably well. We will see later that there are other ways to estimate a functional response model, but the vast majority of functional responses have been estimated using a curve-fitting approach (Uiterwaal et al., 2018). Given the importance of fitting functional response models to the experimental data, and the importance of parameterized functional responses for predicting population dynamics (see Chapter 4), it is important to understand what the models are, where they come from, and how they represent a simplification of the foraging process. The starting point here is the type I functional response.

The *type I functional response* arises owing to (1) a foraging or defensive behavior that does not change with prey density, (2) a predator–prey encounter rate based on random contacts, and (3) the lack of a time cost for the predators to deal with the prey they kill. The type I functional response is analogous to chemical reactions, where the rate of a reaction is proportional to the product of the concentration of the reactants (i.e., mass action). Not all of the reactions that could occur among reactants will occur within a given period, however, because they are distributed in space and it takes time for molecules to come into contact. So this potential amount of reaction is scaled by a reaction rate constant k. The rate of a chemical reaction, A, between reactant R_1 and reactant R_2 is therefore $A = k[R_1][R_2]$, where the brackets indicate concentration. For predator and prey, the mass action term is the product of the predator density [C, consumer; think coyote] and the prey density [R, resource; think roadrunner], or $[C][R]$. Mass action of predator and prey represents literally all the possible encounters between each predator in a population and each potential prey individual. As with reactants, predator and prey are spread out in space, and it takes time for them to come into contact. We therefore scale mass action of predator and prey by some rate constant a, which here I will call the *space clearance rate* (see Box 2.1), and add time spent by the predator searching for prey T_s, to give the total amount of foraging as $F = aCRT_s$, where I have dropped the brackets for simplicity. In the type I functional response, we assume that the only thing a predator does with its time is search for food, so the total time available for a predator (T_{tot}) is equal to its search time (T_s), which means that we can update our equation to $F = aCRT_{tot}$. The functional response describes the per capita (i.e., per individual) rate of foraging, f_{pc} (in this book I will use a small f for foraging rate and a capital F for number of prey individuals consumed). To get to the foraging rate, we divide both sides of the equation $F = aCRT_{tot}$ by T_{tot}, and to get to per capita, we divide both sides of the equation by C. These two steps yield the type I functional response:

$$f_{pc} = aR \tag{2.1}$$

Box 2.1. Why a should be called "space clearance rate."

The functional response parameter a has many names. The terms *attack efficiency*, *attack rate* or *successful attack rate*, *attack constant*, *rate of successful search*, *capture rate*, *maximum clearance rate*, *maximum per capita interaction strength*, *instantaneous rate of discovery*, *rate of potential detection*, and *instantaneous search rate* have all been used as names for this parameter. These names, however, are not good descriptions of the biological process captured by the parameter, as is clearly evident from a unit analysis. The units of the parameter are $\frac{[space]}{[pred][time]}$. We can see this because the units of a line in the functional response space are the units of the rise $\left(\frac{[prey]}{[pred][time]}\right)$ over the units of the run $\left(\frac{[prey]}{[space]}\right)$. These units clearly indicate that the parameter is not a rate of attack, which would be $\frac{[attacks]}{[pred][time]}$. Nor is the parameter an efficiency, which are often unitless, as in the fraction of energy extracted from the energy available in a fuel. Rather, by the units, the parameter a is the space (area or volume) containing prey that is effectively cleared of prey by the predator per unit of time. I formerly preferred the term *area of capture* for this parameter, as this is close to the real meaning and has historical precedent in the *area of discovery* term previously used for parasitoids (Hassell and Varley, 1969). In this book, however, I advocate for renaming this parameter the *space clearance rate*, which is what the units suggest is the biological process being captured.

It is sometimes thought that the type I functional response increases linearly and eventually reaches an abrupt ceiling (occasionally called a "rectilinear" shape), caused by the limits of gut capacity, but this ceiling is rarely observed, difficult to distinguish from a gradual asymptote, and generally ignored in practice (Denny, 2014; Fox, 2013; Jeschke et al., 2004). Jeschke et al. (2004) suggested that in many cases without an observed ceiling (e.g., Monteleone and Duguay, 1988), the functional response also could be viewed as incompletely determined and therefore called a "linear" functional response. Only when higher prey abundances are used would you then be able to determine if the functional response increased linearly to a ceiling or began to taper off to an asymptote.

The *type II functional response* is a modification of type I, where the predator pays a time cost (called a *handling time*) when it captures prey. The handling time includes the time needed for subduing, killing, consuming, and digesting prey and then getting around to searching for more food again. This time is subtracted from the searching time such that the more the predator kills, the less time it searches. The derivation of the type II functional response requires backing up a step from equation (2.1) to the *number* of prey captured per capita (big F_{pc}) in search time T_s:

$$F_{pc} = aRT_s \qquad (2.2)$$

The traditional argument in the derivation of the type II functional response is that predators must divide their time between searching for prey (T_s) and handling prey (T_h), so the total time is $T_{tot} = T_s + T_h$. The handling time cost applies only for prey actually killed (F_{pc}), although it does in practice include the time spent on unsuccessful captures per successful capture. Therefore, the total time budget can be written as $T_{tot} = T_s + F_{pc}h$, where h is the handling time per prey. Since F_{pc} is already spelled out in equation (2.2), we can substitute this into our time budget to get:

$$T_{tot} = T_s + aRT_sh \qquad (2.3)$$

Then we divide the total amount of prey captured by the total time (equation (2.2) for F_{pc} divided by equation (2.3) for T_{tot}) and get:

$$\frac{F_{pc}}{T_{tot}} = f_{pc} = \frac{aRT_s}{T_s + aRT_sh} \qquad (2.4)$$

Factoring out T_s on the right-hand side gives the standard form of the type II functional response:

$$f_{pc} = \frac{aR}{1 + ahR} \tag{2.5}$$

The type II functional response is also called the Holling disc equation, after C.S. Holling, who used students foraging for sandpaper discs to illustrate the shape of the foraging curve. The type II functional response is the most widely used functional response model.

The type II functional response is equivalent to the type I when $h = 0$. Processing prey, however, can never truly take no time—there must always be some time for digestion or managing prey—but that processing time might not always cut into search time. As a result, type I functional responses exist over at least some prey densities (Figure 2.1A), even though particular examples of this type may be just incompletely determined. The derivation of the type II functional response assumes that handling time and searching time are mutually exclusive; that is, predators cannot continue to search for additional prey while they are handling the prey they have already captured. This turns out not to be completely true for all predators. For example, a filter-feeding predator might continue right on filtering even as it clears prey from the water and starts digesting it, so its handling time is zero and its functional response is type I.

The best way to think of the handling time is the loss of searching time owing to handling prey, as only when searching is interrupted does the handling time cause the functional response to saturate, even if handling and searching are not mutually exclusive. Thus, despite the name of the parameter, predators do not have to be actively handling prey to incur handling time, where active handling generally means that the predator is still manipulating, biting, swallowing, or otherwise trying to ingest the prey. They only need to experience a loss of searching time (Abrams, 2000). Different predators may experience losses of searching time in different ways. For example, some predators may be able to digest previous meals while searching for the next one, in which case digestion per se would not be a component of handling time. In other predators, such as some snakes, digestion may be so costly that it could limit additional searching (Secor, 2008). As a result, behavioral identification of handling time as something you can see the predators doing often represents only a portion of the handling time as defined in the derivation of the functional response. Further, since predators are not 100% effective in capturing prey, an estimated handling time from a functional response experiment actually may include an amount of time wasted on unsuccessful attacks per successful attack (Jeschke et al., 2002). Handling prey that are not captured sounds like a violation of the assumptions; however, it can be understood that the handling time for captured prey includes some average number of failed attempts per captured prey.

If we do not factor out the T_s from equation (2.4) as we did earlier, we can write the type II functional response in a way that is more biologically intuitive:

$$f_{pc} = aR \frac{T_s}{T_s + aRT_s h} = aR \frac{T_s}{T_{tot}} \tag{2.6}$$

Equation (2.6) shows that the type II functional response is just the type I model multiplied by the fraction of time actually spent foraging $\left(\frac{T_s}{T_{tot}}\right)$. That is, if a predator spends 50% of its time searching for prey at some prey level, the type II curve at that prey level will be half that predicted by the type I model.

The type II functional response approaches a maximum foraging rate of $1/h$ (asymptote in Figure 2.2A). This should make sense—a predator can only consume one prey per hour if it takes an hour to handle the prey. The space clearance rate can be seen on the functional response curve itself as the slope of the curve as $R \to 0$ (Figure 2.2A). This happens because as R gets very small, the denominator approaches one, meaning that the curve collapses to approximately type I very near the origin, where the equation is just that of a line with a slope equal to the space clearance rate.

The *type III functional response* does not, to my knowledge, have a mechanistic derivation. One simply puts an exponent (such as θ; often called a Hill exponent; Real, 1977) on the prey abundance, R^θ, transforming a rising and saturating curve into one that has a sigmoidal shape (Figure 2.2B):

$$f_{pc} = \frac{aR^\theta}{1 + ahR^\theta} \tag{2.7}$$

The type III functional response may be caused when predators avoid a particular prey type when it is rare, but choose that prey more often when it becomes more common. This prey-switching explanation is the most commonly invoked mechanism for generating a type III functional response, but it is not the only possibility (see Chapters 8 and 9). It is worth pointing out that the type III functional response is often drawn as bending up and reaching the asymptote more slowly than a type II functional response (Denny, 2014). This is not true when one simply adds a Hill exponent to a type II functional

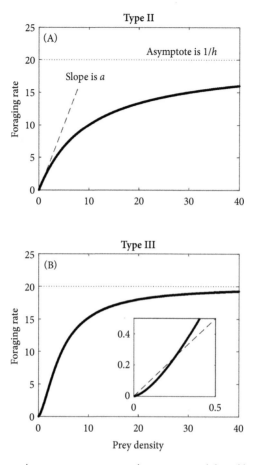

Figure 2.2 How to see the parameters space clearance rate (a) and handling time (h) on a functional response curve. (**A**) For the type II functional response, a is the slope of the curve as it passes through the origin, and the functional response asymptotes at $1/h$. (**B**) For the type III functional response, the asymptote is the same as in the type II but the space clearance rate is not easily visualized along the curve. The inset shows that the curve rises more shallowly than a type II at first but then steepens to rise more quickly. The type III curve also approaches the asymptote more quickly because the effective a ($aR\theta$) is less than a below $R = 1$ and greater than a above $R = 1$.

response while keeping the same space clearance rate, in which case the type III curve crosses the type II curve and actually approaches the asymptote more quickly (Figure 2.2B, inset). It should generally be the case, however, that when estimating a functional response from data, the space clearance rate will be smaller if a type III curve is used as a fitted model rather than a type II curve (see Chapter 9).

As with the type II functional response, the saturation point on a type III functional response is $1/h$ (Figure 2.2B). The location of the space clearance rate on a type III functional response is much harder to visualize than on the type II (see Chapter 9). It is helpful to realize that the Hill exponent is really just a way to modify a such that predators do not clear much space when that specific prey type is rare. That is, a can be a function of R (Juliano, 2001). The typical way to do this is to rewrite a as an increasing function of prey density, such as $a = bR^q$, where $q > 0$ and the Hill exponent $\theta = q + 1$, and substitute this into equation (2.7):

$$f_{pc} = \frac{(bR^q)\,R}{1 + (bR^q)\,hR} \tag{2.8}$$

Both the θ (equation (2.7)) and q (equation (2.8)) versions of the type III functional response are used in the literature (Smout et al., 2010; Vucic-Pestic et al., 2010).

The *type IV functional response* is the case when foraging rate declines at high densities. That is, after prey levels have increased high enough for the predator to approach the asymptotic foraging rate, further increases in prey levels cause the foraging rate to decline again (Figure 2.1D). This effect could be caused by predator confusion in the presence of a lot of prey or by the risk to the predators caused by an increase in the number of dangerous prey (Jeschke and Tollrian, 2005). This type of functional response appears to be quite rare, and some have even suggested that it should be viewed as a mathematical artifact (Morozov and Petrovskii, 2013). However, it has been seen for the odonate predator *Aeshna cyanea* foraging on the cladoceran *Daphnia magna* (Jeschke and Tollrian, 2005) and *Zodarion* spiders foraging on a variety of ants, which may be related to the danger ants pose to their predators in larger groups (Líznarová and Pekár, 2013). One way to model this would be to propose that handling time increases as prey levels increase, perhaps because the handling challenge is more difficult when faced with the risk incurred from the presence of numerous uneaten prey. Thus, we could simply add another exponent to prey levels in the denominator of the type II functional response:

$$f_{pc} = \frac{aR}{1 + ahR^g} \tag{2.9}$$

As with the Hill exponent, here the g is just a phenomenological parameter generating a particular effect rather than an easily interpretable biological process.

2.2 **Predator dependence of the functional response**

All of the above functional response types are curves that link foraging rate to one variable—prey density. Per capita foraging rates, however, usually also decrease with the density of predators (Figure 2.3). This effect is often called *interference* because each predator has a negative effect on the foraging rate of all the other predators (DeLong and Vasseur, 2011; Hassell, 1971; Skalski and Gilliam, 2001). The type II functional response can be modified in several ways to accommodate interference competition (Novak and Stouffer, 2020; Skalski and Gilliam, 2001; Tyutyunov et al., 2008). One common way is to raise predator density to a power, called *m* for *mutual interference*, in much the same way that θ was added to the functional response to create a type III functional response (this is the Hassell–Varley form of interference) (Arditi et al., 1991; Hassell, 1971):

$$f_{pc} = \frac{aRC^m}{1 + ahRC^m} \qquad (2.10)$$

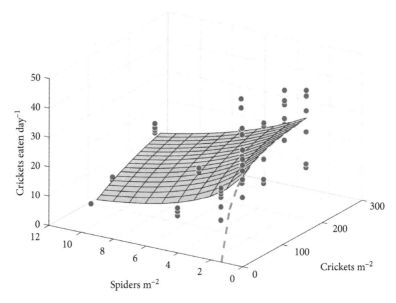

Figure 2.3 The foraging rate of the wolf spider *Pardosa milvina* foraging on the cricket *Acheta domesticus* in response to both prey and predator density, replotted from the data in Schmidt et al. (2014) and fit to equation (2.10). A saturating type II response can be seen along the cricket axis where there is only one spider (dashed line), but the rate of foraging declines at all prey densities as spider density increases, which is known as predator interference.

The mutual interference exponent m is negative to indicate that adding more predators reduces the foraging rate, but in principle the exponent could be positive to indicate some mutually beneficial effect of more predators on the average foraging rate, such as group foraging. As with the type III functional response, what is happening here is that a is being made a decreasing function of predator density, rather than an increasing function of prey density as in the type III functional response:

$$f_{pc} = \frac{(aC^m)\,R}{1 + (aC^m)\,hR} \tag{2.11}$$

The a in equation (2.10/2.11) is the same as the a in equation (2.5), because equation (2.5) is just the special case of equation (2.10/2.11) when $C = 1$. When $m = -1$, the functional response is dependent on the ratio of prey to predator abundance, which is known as *ratio-dependent* (Akcakaya et al., 1995). When $m = 0$, the functional response is *prey-dependent*, meaning only dependent on prey levels and not dependent on predator levels. The ratio-dependence idea stimulated some controversy about how important predator density is to functional responses; for more on this, see Chapter 3.

Another common way of including interference in a type II functional response is to add a "wasted time" term to the time budget (Beddington, 1975; DeAngelis et al., 1975). This approach (the Beddington–DeAngelis form) has a clearer derivation, where you simply add the amount of time wasted for each predator–predator interaction. Specifically, we add the time wasted per predator, w, times the number of predators minus one (so that there is no interference when there is only one predator) to the time budget in equation (2.3) and follow the same steps until you get:

$$f_{pc} = \frac{aR}{1 + w\,(C - 1) + ahR} \tag{2.12}$$

These two models for predator interference (equations (2.11) and (2.12)) differ not just mathematically but in what the functions imply about the mechanism of interference. In the Hassell–Varley expression (equation (2.11)), interference is caused by a reduction of the space clearance rate, which might be caused by several behavioral changes in the predator that influence their ability to move in space or acquire prey. In the Beddington–DeAngelis expression (equation (2.12)), interference is caused by a loss of searching time, meaning a change in the predator's time budget. When fitted to data, both models look very similar in practice, which may not be surprising because both models describe the outcome of some cost to the predator from interacting with other predators (Kratina et al., 2009; Skalski and Gilliam, 2001). More detailed behavioral analyses are required to identify the mechanisms generating predator interference and to guide the use of specific interference models for any particular system. There is also no reason to assume that all cases of interference are generated by the same mechanism, so the two models presented here, or other models, might still be required to properly describe the effect of predator abundance on foraging rates across the different predators in food webs (Crowley and Martin, 1989; DeLong and Vasseur, 2013; Skalski and Gilliam, 2001; Tyutyunov et al., 2008).

2.3 Relationship to alternative formulations in aquatic literature

Many studies focused on aquatic invertebrates or microbes, as well as some theoretical studies, use an alternative formulation of the functional response known as a Michaelis–Menten equation:

$$f_{pc} = \frac{I_{max}R}{K_R + R} \tag{2.13}$$

where I_{max} is the maximum ingestion rate and K_R is the half-saturation constant. This constant is the prey level at which the foraging rate reaches half of its maximum value I_{max}. What happens with this equation is that as R gets very large, the relative effect of adding K_R to R in the denominator gets small. In other words, $\frac{R}{K_R+R} \to 1$ as $R \to \infty$. Thus, the asymptote is I_{max}, which means $I_{max} = 1/h$, just as before. You also can transform equation (2.5) into equation (2.13) by multiplying the top and bottom by $1/ah$ (Fan and Petitt, 1994):

$$f_{pc} = \frac{aR}{1 + ahR} \frac{1/ah}{1/ah} = \frac{\frac{1}{h}R}{1/ah + R} = \frac{I_{max}R}{K_R + R} \tag{2.14}$$

which also clarifies that the half-saturation constant $K_R = 1/ah$. We will not use this formulation in this book except once at the very end, because the combined parameter K_R reflects both searching and handling processes and is therefore more difficult to understand and interpret biologically. Nonetheless, the form continues to be used in the literature, perhaps because of the convenience of having a maximum foraging rate parameter in the model.

Another common rate in the aquatic literature is the "clearance rate," which is actually a little different than space clearance rate. This is calculated as the foraging rate f_{pc} divided by the prey levels. Returning to equation (2.6), we can see that the clearance rate in the aquatic literature is equal to the product of space clearance rate and the fraction of time spent searching:

$$C = \frac{aR\frac{T_s}{T_{tot}}}{R} = a\frac{T_s}{T_{tot}} \tag{2.15}$$

Thus, the difference between the *space clearance rate* that is a component of the functional response (a) and the *clearance rate* in the aquatic literature is that the former is a constant and the latter captures the effect of the time spent handling (which increases with more prey) on the space clearance rate.

2.4 The Rogers Random Predator equation

An important alternative expression for the functional response arises from accounting for the common experimental artifact of declining prey levels during functional response experiments. A typical functional response experimental set-up has an arena within which predators forage on some amount of prey. As the predators forage, one could replace each consumed prey item with a fresh prey item, thus keeping the prey levels constant. Doing this can be quite disruptive to the predator or impractical, so in most cases the prey simply decline in abundance through time as the predators consume them. That means that with each prey captured, the expected foraging rate on the new prey level declines, reducing the apparent foraging rate for the original prey density. We can account for this declining prey density using what is known as the Roger's Random Predator equation (RRP; see Box 2.2 for the derivation of equation (2.16); Rogers, 1972; Royama, 1971):

$$R_e = R_o\left(1 - e^{a(hR_e - t)}\right) \tag{2.16}$$

In this equation, the foraging rates have been replaced by numbers of prey offered (R_o) and eaten (R_e), while a and h are still space clearance rate and handling time, respectively, and the duration of the experiment is t. However, because the RRP generally is used with the number of prey items offered, not the number per unit area, the default units of space clearance rate in this context are arenas (the physical space of the experimental area, regardless of whether the arena is a surface or a volume) per predator per time.

Box 2.2. Derivation of the Rogers Random Predator equation.

We start with a differential equation that describes the rate of change in prey abundance (R) through time (t) owing to predation:

$$\frac{dR}{dt} = -\frac{aR}{1 + ahR}$$

This equation describes a rate of prey loss owing to predation described by a type II functional response. If we integrate the equation over the course of an experiment of duration t, we will get the cumulative loss of prey from the amount of prey offered at the beginning, R_o, to the number of prey items left at the end of the experiment, R_t. First we arrange the equation to aggregate R to the left-hand side and time to the right-hand side:

$$\frac{(1 + ahR)}{R} dR = -adt$$

and take the integrals from R_o to R_t and from 0 to t on the left-hand side and right-hand side, respectively:

$$\int_{R_o}^{R_t} \frac{(1 + ahR)}{R} dR = \int_{R_o}^{R_t} \frac{1}{R} dR + \int_{R_o}^{R_t} ahdR = \int_0^t -adt$$

Integration yields:

$$\ln R_t - \ln R_o + ahR_t - ahR_o = -at$$

which we can rearrange as:

$$\ln \frac{R_t}{R_o} = -at - ahR_t + ahR_o$$

and then further to:

$$\ln \frac{R_t}{R_o} = a(-t - hR_t + hR_o)$$

Finally, exponentiating both sides followed by moving the R_o to the right-hand side gives:

$$R_t = R_o e^{a(h(R_o - R_t) - t)}$$

There are two tricks needed to get to the final equation. First, subtract each side from both sides (i.e., switch sides and change sign):

$$-R_o e^{a(h(R_0-R_t)-t)} = -R_t$$

Second, recognize that the number of prey actually eaten is the difference between the number offered and the number that is left: $R_e = R_o - R_t$. We therefore add R_o to both sides:

$$R_o - R_o e^{a(h(R_o-R_t)-t)} = R_o - R_t$$

Since the right-hand side is now the number of prey eaten, we just substitute and factor out the R_o, bringing us to equation (2.16):

$$R_e = R_o \left(1 - e^{a(hR_e-t)}\right)$$

Because R_e occurs on both the left-hand side and right-hand side of equation (2.16), it needs to be solved using numerical methods (i.e., searching for combinations of parameters that satisfy the equation). More recently, Bolker (2011) showed that the RRP can be simplified to isolate the R_e on the left-hand side and make fitting of the equation to functional response data much easier:

$$R_e = R_o - \frac{W\left(ahR_0 e^{-a(t-hR_0)}\right)}{ah} \tag{2.17}$$

where W is the LambertW (or product log) function (see Box 2.3 for the derivation of equation (2.17)). I will call equation (2.17) the Lambert Random Predator equation. Remembering that a type III functional response is one in which space clearance rate is a function of prey level, $a = bR^q$, it might seem that we can substitute this term into equation (2.17) to get the equivalent type III formulation:

$$R_e = R_o - \frac{W\left((bR^q) hR_0 e^{-(bR^q)(t-hR_0)}\right)}{(bR^q)h} \tag{2.18}$$

However, this is just an approximation and may produce slightly biased estimates of the foraging rates (Bolker, 2011; Rosenbaum and Rall, 2018). For more on type III functional responses, see Chapter 9.

Box 2.3. Derivation of the Lambert$W$ version of the Rogers Random Predator equation.

Bolker (2011) starts with the original RRP (see text and Box 2.2) and does a bunch of clever rearranging to get to a solution. Since only part of those steps is shown in the original work, I will write out all of the steps here so I can remember them. Starting with the original RRP:

$$R_e = R_o \left(1 - e^{a(hR_e - t)}\right)$$

we can rearrange to get:

$$1 - \frac{R_e}{R_o} = e^{-a(t - hR_e)}$$

Expand the exponential term:

$$1 - \frac{R_e}{R_o} = e^{-at + ahR_e} = e^{-at}e^{ahR_e}$$

The first trick is to multiply the right-hand side by $\frac{e^{ahR_o}}{e^{ahR_o}}$ and combine the first exponential term with the numerator and the second with the denominator:

$$1 - \frac{R_e}{R_o} = \left(e^{-at}e^{ahR_o}\right)\left(\frac{e^{ahR_e}}{e^{ahR_o}}\right)$$

Now follow exponential rules and factor out:

$$1 - \frac{R_e}{R_o} = \left(e^{-a(t - hR_o)}\right)\left(e^{ah(R_e - R_o)}\right)$$

The second trick is to multiply the exponent in the second exponential term by $\frac{R_o}{R_o}$, which allows us to rearrange to:

$$1 - \frac{R_e}{R_o} = \left(e^{-a(t - hR_o)}\right)\left(e^{ahR_o\left(\frac{R_e}{R} - 1\right)}\right)$$

The third trick is to multiply both sides by ahR_o and move the second exponential term to the left-hand side:

$$\left[ahR_o\left(1 - \frac{R_e}{R_o}\right)\right]e^{\left[ahR_o\left(1 - \frac{R_e}{R_o}\right)\right]} = ahR_o\left(e^{-a(t - hR_o)}\right)$$

Bolker did all these tricks to get to this special place. Once here, he offered a clever solution to this arrangement by noticing that it takes the form of the product log, which is that for some function $y = xe^x$, the solution is $W(y) = x$. In this example, $x = ahR_o \left(1 - \frac{R_e}{R_o}\right)$, and this means that the last equation can be written as:

$$ahR_o \left(1 - \frac{R_e}{R_o}\right) = W\left(ahR_o \left(e^{-a(t-hR_o)}\right)\right)$$

Now rewrite as:

$$ah\left(R_o - R_e\right) = W\left(ahR_o \left(e^{-a(t-hR_o)}\right)\right)$$

And with just a little more rearranging we get to equation (2.17):

$$R_e = R_o - \frac{W\left(ahR_o \left(e^{-a(t-hR_o)}\right)\right)}{ah}$$

3

What Causes Variation in Functional Response Parameters?

The parameters of the functional response are not traits. They represent processes such as hunting and digesting prey. Thus, all the traits that influence the way predators and prey encounter each other in space and the morphologies and behaviors that influence capture, evasion, or digestion are potential sources of variation in the functional response parameters. In this chapter, I cover how we break the parameters down mathematically so that the connection between the parameters and traits is more transparent.

3.1 Variation in functional response parameters

Functional response parameters can take on a wide range of values (Figure 3.1). Across all kinds of predator and prey pairs represented in the FoRAGE database (**F**uncti**o**nal **R**esponses from **A**round the **G**lobe in all **E**cosystems), variation in space clearance rate exceeds 15 orders of magnitude, while variation in handling time exceeds 8 orders of magnitude. This massive amount of variation in functional response parameters stems from variation among individual predators as well as variation across pairs of predator and prey species. Where does such variation come from?

Among the many possible sources of variation, body size of both predator and prey seems to play a large role. There has been tremendous effort to quantify how predator and prey body sizes influence functional response parameters, both within and across species (Byström and García-Berthou, 1999; DeLong and Vasseur, 2012a; Gergs and Ratte, 2009; McCoy et al., 2011; Miller et al., 1992; Rall et al., 2011, 2012; Rindorf and Gislason, 2005; Schröder et al., 2016; Spitze, 1985; Uiterwaal et al., 2017; Uiterwaal and DeLong, 2018; Vucic-Pestic et al., 2010; Weterings et al., 2015). Even when accounting for body size, however, substantial variation in functional response parameters remains. This additional variation can be linked to taxonomic identity

Predator Ecology: Evolutionary Ecology of the Functional Response. John P. DeLong, Oxford University Press.
© John P. DeLong 2021. DOI: 10.1093/oso/9780192895509.003.0003

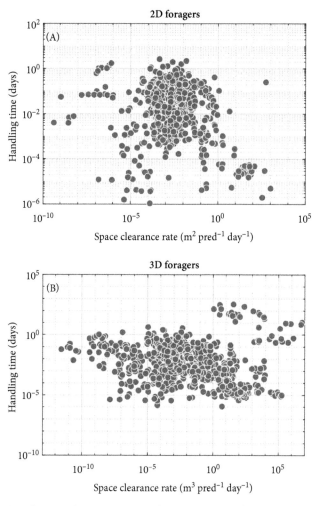

Figure 3.1 Pairs of space clearance rate and handling times for >2,000 predator–prey pairs of all types of species from around the world. Data are from the global compilation of functional responses called FoRAGE (Uiterwaal et al., 2018). Because the units of space clearance rate depend on the actual type of space being used, the data are plotted separately for two-dimensional interactions (**A**) and three-dimensional interactions (**B**).

(Kalinoski and DeLong, 2016; Rall et al., 2012; Uiterwaal and DeLong, 2018, 2020), temperature (Bergman, 1987; Dell et al., 2014; Englund et al., 2011; Rall et al., 2012; Uiterwaal and DeLong, 2020; Uszko et al., 2017), age, or instar, which also may reflect variation in body size or energetic demands (Gergs and Ratte, 2009; Hassanpour et al., 2011; Houde and Schekter, 1980; McArdle and Lawton, 1979; McCoy et al., 2011; Spitze, 1985; Uiterwaal and

DeLong, 2018; van den Bosch and Santer, 1993; Yaar and Özger, 2005), sex (Walker and Rypstra, 2002), habitat complexity (Alexander et al., 2015; Barrios-O'Neill et al., 2015; Cuthbert et al., 2019; Koski and Johnson, 2002; Kreuzinger-Janik et al., 2019; Nilsen et al., 2009; Ware, 1972), aggressiveness or choosiness about prey types (Michalko et al., 2017), individual variation in dominance (Hartley et al., 2019), and nutritional state or energetic condition (i.e., "hunger") (Li et al., 2018; Lyon et al., 2018; McMahon and Rigler, 1965; Nandini and Sarma, 1999; Schmidt et al., 2012).

Understanding how this variation arises and how specific traits or processes mechanistically influence functional response parameters requires some dissection of the processes that generate predation events in the first place. As with most model parameters in biology, the parameters of the functional response are composed of other parameters. For example, the intrinsic rate of population growth is a parameter r, but it is also the difference between two other parameters—the rate of births b and the rate of deaths d. Thus, to understand the space clearance rate and the handling time parameters, we must take them apart and look at what other parameters combine to produce them. Breaking down the parameters into their component parts will help us understand the foraging mechanisms that lead to specific functional response parameter values, why there is so much variation in parameters, and how evolution that acts on traits linked to foraging may be mediated by the consequences of the functional response (Denny, 2014; Jeschke et al., 2002; Livdahl, 1979).

3.2 Breaking down the space clearance rate

The space clearance rate is the fundamental foraging parameter, as it determines how fast predators can forage when they pay no time costs for handling prey. Successful foraging requires the predator to encounter, attack, and successfully kill prey (Endler, 1986; Gentleman et al., 2003; Roberts et al., 2010; Sih and Christensen, 2001). We can use these three steps to break down the space clearance rate into its component parameters. First, encounters arise owing to the movement of predator and prey in space and are thus proportional to the root sum of square of the velocity of the predator (V_c) and the prey (V_r): $\sqrt{V_c^2 + V_r^2}$ (Aljetlawi et al., 2004; Pawar et al., 2012) (for derivation of the root sum of squares expression, see Box 3.1). The rate of encounters is also proportional to the amount of prey in the environment, so as either the predator or the prey increase in speed, or prey become

---

Box 3.1. Why the root sum of squares?

The root sum of squares expression ($\sqrt{V_c^2 + V_r^2}$) arises from recognizing that the relative velocities ($\vec{V}_{relative}$) of two objects determine the rate at which they collide. Going back to a classic math problem as an example, the relative velocity of one car passing another is the difference in their velocities. That is, the fast car is going past the slow car at relative velocity $\vec{V}_{relative} = \vec{V}_{fast} - \vec{V}_{slow}$. If the fast car was not moving, the slow car would be moving away at its own velocity, and if the slow car stopped moving, the fast car would pass it at its own velocity. By analogy, gas molecule-like predators (C) and prey (R) move past each other at a relative velocity of $\vec{V}_{relative} = \vec{V}_C - \vec{V}_R$. The above root sum of squares expression comes from taking the average of the relative velocities of all the predator and prey individuals (Pawar et al., 2012). Because both predators and prey are on average not moving anywhere (i.e., the displacement of all the different individuals in different directions sums to zero), we can ignore their starting points and focus on the mean relative velocity. And because the angle separating their velocities is as likely to be 90° as anything else, we can use the Pythagorean Theorem to calculate the average relative velocity as the hypotenuse of the triangle formed by the two velocities moving away from each other from the same point and at 90°:

$$\vec{V}_{relative} = \sqrt{\vec{V}_C^2 + \vec{V}_R^2}$$

which is written in equation (3.2) without the arrows for convenience. Note that key features of different kinds of predators correctly map onto this expression. The relative velocity of a sit-and-wait predator is that of the prey, and the relative velocity of a predator foraging on sessile prey is that of the predator. That outcome would not occur if we used the arithmetic mean of the two velocities.

---

more abundant, the encounters between predators and prey will increase. Increasing predator abundance also increases encounters, but remember the functional response is a per capita expression. Second, an actual encounter arises only if predators are aware of the prey item, so we include a term for the distance, d, at which a predator can detect prey. Detection will be a length if the predator is searching in two dimensions and an area for predators searching in three dimensions (Pawar et al., 2012). Importantly, this detection distance is the distance that a predator can perceive a prey, not the maximum distance that it can perceive anything. For example, a lion may be able to see a zebra at a great distance, but it may not be able to distinguish it from the tall grass until it is much closer. Third, after encountering the prey, the predator must choose to attack the prey, and it does so with some probability, p_a. Finally, the predator has some chance of actually being successful given that it has attacked the prey, p_s. Putting this together, with the events transpiring from right to left, the space clearance rate is:

$$a = p_s p_a d \sqrt{V_c^2 + V_r^2} \qquad (3.1)$$

Equation (3.1) shows how the foraging process leads to the emergent parameter space clearance rate, but the equation also helps to clarify why the units of space clearance rate come out as $\frac{[space]}{[pred][time]}$. The units of the velocities are $\frac{[length]}{[time]}$ (the distance traveled by an individual predator or prey in a unit of time), the detection distance is either $\frac{[length]}{[pred]}$ for a two-dimensional forager or $\frac{[length]^2}{[pred]}$ for a three-dimensional forager, and the probabilities are unitless. Multiplying them together as in equation (3.1) makes the units of space clearance rate $\frac{[area]}{[pred][time]}$ for a two-dimensional forager or $\frac{[volume]}{[pred][time]}$ for a three-dimensional forager. We can then generalize the units of the numerator as "space." It is worth noting here that some sources suggest the space clearance rate represents a preference for a particular prey type (Heong et al., 1991; Smout et al., 2013). Preference for a prey type may influence space clearance rate by influencing the probability of attack, but the space clearance rate itself reflects preferences along with all of the other factors influencing predation events, so space clearance rate should not be thought of as only reflecting preference.

Equation (3.1) also can help clarify why space clearance rate, encounter rate, attack rate, and foraging rate are not the same things, even though many papers refer to space clearance rate as "attack rate" (Figure 3.2). Space clearance rate is a property of the interaction between predator and prey that tells us how effectively a predator clears searched space of their prey. Successful foraging depends on encounters between predators and prey, which is the product of prey density and the rate at which space is perused for prey, so encounter rate $E = d \sqrt{V_c^2 + V_r^2} R$. Some portion of those encountered prey are attacked, giving the actual attack rate as $A = p_a d \sqrt{V_c^2 + V_r^2} R$. Some portion of

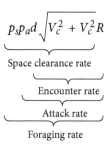

Figure 3.2 How the parameters involved in determining the space clearance rate are repackaged into encounter rate, attack rate, and foraging rate.

these attacked prey are successfully predated, giving the total foraging rate for a type I functional response (with no handling time) as:

$$f_{pc} = p_s p_a d \sqrt{V_c^2 + V_r^2} R = aR \tag{3.2}$$

One of the useful things about equation (3.2) is that it is now easy to imagine specific traits of predators or prey that could influence the foraging steps and thus influence the parameter itself. Traits associated with speed of movement, such as limb lengths, swimming ability, activity level, or body size, can influence encounter rates. For example, smaller crabs (*Panopeus herbstii*) foraging on mussels (*Brachidontes exustus*) that were more active had a higher functional response than less active crabs, presumably through an effect on searching velocity (Toscano and Griffen, 2014). Traits associated with information gathering such as sight, smell, and the ability to sense vibrations can all influence the detection distance. The probabilities of attack can be linked to a predator's interest in a type of prey, set possibly by potential energetic or nutritional reward as food or the cost to the predator through prey defenses such as toxins or spines. The probability of capture success could likewise be influenced by traits that facilitate or disrupt capture, such as acceleration in the predator or prey, grasping structures, or again, prey defenses.

3.3 Factors affecting space clearance rate

All foraging activity occurs in physical space characterized by habitat and dimensionality (e.g., a flat plane, the branching structure of a plant, the open ocean, or the air). Functional responses therefore reflect the structure and complexity of the foraging habitat that constrains or facilitates the foraging process by determining where organisms can be or how they can move. If habitat complexity alters movement patterns or the ability to detect or avoid predators, then the space clearance rate should reflect this. For example, kokanee salmon (*Oncorhynchus nerka*) showed a steeper functional response foraging for *Daphnia* in high light than in low light, presumably owing to greater prey detectability (d) in brighter conditions (Koski and Johnson, 2002), and lynx showed steeper functional responses in winter than in summer, potentially because of the difficulty of prey escape during winter (an effect on p_s) (Nilsen et al., 2009).

Habitat complexity can reduce the encounters among predators and prey and therefore can make a functional response shallower for some predators (Alexander et al., 2015; Barrios-O'Neill et al., 2015; Cuthbert et al., 2019; Ware, 1972). Moreover, increasing habitat complexity can generate the potential for refuges, altering not just the height of the functional response but its shape. In the case of dragonfly larvae (*Anax junius*) foraging on tadpoles (*Rana pipiens*), the functional response shifted to type III in the presence of prey refugia, suggesting reduced encounters between predator and prey particularly at low prey densities (Hossie and Murray, 2010). In other systems, however, predators may show an increase in space clearance rate with habitat complexity (Wasserman et al., 2016). In the case of the notonectid predator *Enithares sobria* foraging on the water flea *Daphnia longispina*, the effect of habitat complexity depended on temperature, suggesting that the physical aspects of predator or prey movement interacted with the physiological effects of temperature on predator, prey, or both (Figure 3.3; Wasserman et al., 2016). Similarly, increasing the complexity of arena edges can reduce functional responses by creating refugia (Hoddle, 2003), but increasing the area of arenas can increase the functional response by concentrating prey along the susceptible outer edge making prey more vulnerable (Uiterwaal et al., 2019).

In addition to habitat complexity, predators that forage in groups may have higher detection, more encounters, or greater probability of success, increasing the functional response relative to individual foraging (Fryxell et al., 2007; Nilsen et al., 2009). Also, "hunger" levels, as determined by how much a predator has recently fed, would logically influence space clearance rate, since the decision to attack (p_a) or search velocity (V_c) should be motivated by how much the predators need to eat (Charnov, 1976; Jeschke, 2007; Jeschke et al., 2002). This expectation seems to be met in *D. magna* foraging on a range of food types (McMahon and Rigler, 1965) and for *Hogna baltimoriana* wolf spiders foraging for grasshoppers (Lyon et al., 2018).

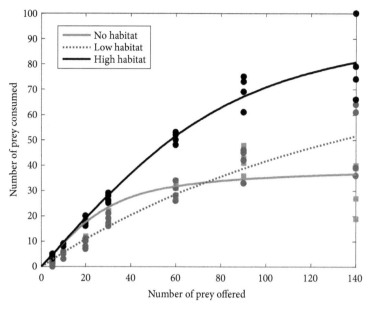

Figure 3.3 Functional responses of the notonectid predator *Enithares sobria* foraging on the cladoceran *Daphnia longispina* at 14°C. Data from Wasserman et al. (2016). Functional responses were lower for both no and low levels of habitat complexity (zero and two plant stalks in the arenas) than with high levels of habitat complexity (four stalks).

As already indicated, one of the better-investigated traits linked to functional response parameters is body size (Goss-Custard et al., 2006; MacNulty et al., 2009; McCoy et al., 2011; Miller et al., 1992; Thompson, 1975). Because body size is often positively associated with the speed at which both predators and prey move through the environment through a power-law-like relationship (Calder, 1996; Peters, 1983), we might predict that space clearance rate has a power-law-like relationship with either predator and/or prey body size (Aljetlawi et al., 2004; McGill and Mittelbach, 2006) (Figure 3.4). This kind of relationship has been found across species for several types of organisms, including mammalian carnivores (DeLong and Vasseur, 2012a), protists (DeLong and Vasseur, 2012b; Fenchel, 1980a, b), ladybird beetles (Uiterwaal and DeLong, 2018), and across multiple taxa together (Rall et al., 2012; Uiterwaal and DeLong, 2020). Although there is an apparent "universal" scaling across all species (Figure 3.4A), there are distinct scalings for some taxonomic groups (Figure 3.4B; Uiterwaal and DeLong, 2020). The within-group scalings also may be shallower than the overall scaling, such that the overall scaling intersects multiple groups and obscures some potentially important variation. The body size-dependence of

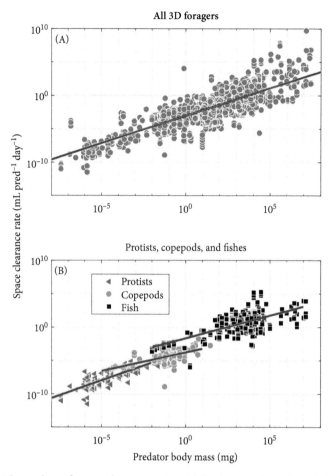

Figure 3.4 The scaling of space clearance rate with body mass, based on data from the FoRAGE database. (**A**) All the three-dimensional foragers in the database together scale strongly with body mass, with an exponent of 0.82, not considering other factors. (**B**) Breaking these foragers into three taxonomic groups that also tend to vary in body size, it is clear that the overall scaling crosses some group-specific scaling patterns, with the scaling exponents somewhat shallower for fishes (0.68) and copepods (0.56) but in line with that for protists (0.89).

functional responses both within and across species may even be strong enough to negate species-level differences, such as was shown for some freshwater fishes foraging on brine shrimp (Miller et al., 1992). Interestingly, because of the different power-law scalings of velocity for two-dimensional and three-dimensional foragers, it also has been shown that the exponent for the scaling of space clearance rate on body size is steeper for three-dimensional foragers than two-dimensional foragers (Pawar et al., 2012).

Body size also influences space clearance rate within predator–prey pairs. For example, the space clearance rate of individual least killifish (*Heterandria formosa*) foraging on brine shrimp (*Artemia salina*) was positively correlated with predator body size (Schröder et al., 2016). In other cases, variation is associated with prey body size. Space clearance rates got smaller as prey body size increased for both water bugs (*Belostoma* sp.) and dragonfly nymphs (*Pantala flavescens*) foraging on red-eyed treefrog tadpoles (*Agalychnis callidryas*) (McCoy et al., 2011). This effect could arise through increased escape speeds for larger prey, affecting the probability of success (p_s) (Gvoždík and Smolinský, 2015). Sometimes this body size effect is linked to age or instar of the predator. For example, backswimmers (*Notonecta maculata*) are aquatic insects that grow through multiple instars on their way to adulthood, and they show clear variation in their functional response owing to age and body size. Gergs and Ratte (2009) showed that as backswimmers grow through these instars, their functional response for foraging on water fleas (*D. magna*) changes (Figure 3.5). The largest backswimmer instars have a functional response that increases with water flea size but declines again for the largest water fleas (Figure 3.5A). This pattern indicates that the space clearance rate peaks at intermediate prey sizes (Figure 3.5B). This peak, however, is lower for the smaller backswimmers and higher for the larger backswimmers.

In the backswimmers, age and body size have strong effects on the functional response, creating variation in how strongly specific backswimmers interact with specific water fleas. Spitze (1985) found a similar pattern for phantom midges (*Chaoborus americanus*) foraging on a different species of water flea (*Daphnia pulex*). Why does such variation occur? Equation (3.1) indicates it could arise through several avenues, but Gergs and Ratte (2009) showed that encounter rates increased with water flea size (an effect on V_c), larger backswimmers detected water fleas from farther away (an effect on d), and the probability of success (p_s) peaked at intermediate backswimmer size. In a different study, Streams (1994) found that encounter rates and the probability of attack (p_a) both increased with body size of backswimmers (*Notonecta undulata*), but the proportion of successful attacks (p_s) decreased with predator body size. Thus, several aspects of the foraging process related to body size, movement, and traits interact to generate variation in the functional response.

Equation (3.1) shows that it is the product of multiple events that generates space clearance rate. Thus, if some factors increase while others decrease with some trait, it is not difficult for space clearance rate to peak at intermediate values of that trait, as in the case with backswimmer body size. Given the multiple ways in which predator and prey body size influence the components

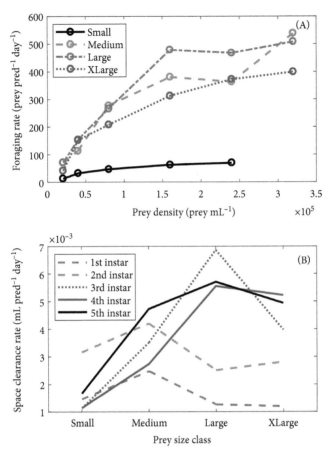

Figure 3.5 Functional responses of backswimmers (*Notonecta maculata*) foraging on *Daphnia magna*. Data from Gergs and Ratte (2009). (**A**) Functional responses vary with daphnid size, with the steepest and highest curves for the largest daphnids. (**B**) The space clearance rate of backswimmers varies with both backswimmer instar and daphnid size. There is a clear peak in each curve, with smaller backswimmers showing a peak in space clearance rate on medium-sized daphnids and larger backswimmers showing a peak with the larger daphnids.

of foraging, including encounters, morphologies such as gape limitations, and prey defenses, it may not be too surprising that several studies have documented a peaked relationship between predator body size or predator–prey body size ratio and space clearance rate (Barrios-O'Neill et al., 2016; Cuthbert et al., 2020b; Gergs and Ratte, 2009; Osenberg and Mittelbach, 1989; Spitze, 1985; Uiterwaal et al., 2017; Vucic-Pestic et al., 2010). Such peaks sometimes are thought of as a kind of optimal prey size for the predator, but the peaks may or may not represent optima in terms of predator fitness. Also, although it is common to think of the predator–prey body size ratio as a good predictor of space clearance rate, it may be that predator and prey body size may explain more variation when taken as independent factors. For example, across all predator–prey pairs in the FoRAGE database, a model including predator-prey body size ratios was effective in explaining variation in space clearance rate, but a model with predator and prey body size included as independent factors was more supported by the data (Uiterwaal and DeLong, 2020). This outcome makes sense, as the ratio forces a specific type of interaction between predator and prey body mass effects as a predictor, but using predator and prey body sizes as independent factors allows them to explain the data more flexibly. This outcome is also what we would expect if predator and prey body size affected the foraging process in different ways.

The effect of mass may not always be very strong, as seen in freshwater cyclopoid copepods, where the presence of a hard outer shell seems more important that just prey body size (Kalinoski and DeLong, 2016). In addition to fixed defenses, inducible defenses, a form of phenotypic plasticity where in the presence of predators, new prey behaviors or morphologies emerge (either quickly or across generations) that can alter the functional response by reducing some component of space clearance rate. For example, the mosquito larvae *Culex jenningsi* showed reduced movements and relocated to safer surface waters in the presence of chemical signals from predatory damselfly nymphs (Hammill et al., 2015). Presumably, these changes reduced encounters, detectability, and perhaps the probability of success for the predatory damselflies. Similarly, *D. pulex* may develop an induced structural defense (neckteeth) in response to foraging by the phantom midge *Chaoborus obscuripes* that reduces the height of the functional response (Jeschke and Tollrian, 2000). In contrast, some predators can display phenotypic plasticity, such as with overall body size, as an inducible offense in response to prey (Kopp and Tollrian, 2003). Several studies have found that invasive predators have higher functional responses on naïve native prey than similar native predators, suggesting that co-evolved defenses that are effective against native predators may not work well against novel invasive predators (Alexander et al., 2014; Bovy et al., 2015; Crookes et al., 2019; Kiesecker and Blaustein, 1997).

The components of space clearance rate respond to temperature. In particular, the searching velocities of ectotherm predators and prey may depend on temperature, if they are not sit-and-wait predators, and this temperature dependence of velocity may be reflected in the temperature dependence of space clearance rate (Dell et al., 2014). Consistent with this idea, space clearance rate often increases with temperature across many species

(Burnside et al., 2014; Kalinoski and DeLong, 2016; Rall et al., 2012; Uiterwaal and DeLong, 2018). Space clearance rate may increase with temperature within predator–prey pairs as well (Thompson, 1978). A study with the ladybird beetle *Harmonia axyridis* foraging on cutworm moth (*Spodoptera litura*) eggs showed an increase in space clearance rate with temperature, but this temperature effect varied with predator instar (Islam et al., 2020). A unimodal relationship between space clearance rate and temperature often emerges within predator–prey pairs if a wide enough temperature range is evaluated (Sentis et al., 2012; Song and Heong, 1997; Uszko et al., 2017). For example, the functional response of thrips (*Scolothrips takahashii*) foraging on hawthorn spider mites (*Tetranychus viennensis*) overall increased in height with temperature for both females (Figure 3.6A) and males (Figure 3.6B) (Ding-Xu et al., 2007). Despite the general increase, however, space clearance rate may have peaked at an intermediate temperature (Figure 3.6C). Most surprising, the peak in space clearance rate was at a higher temperature for

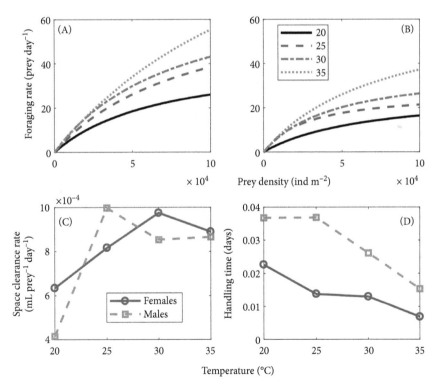

Figure 3.6 Functional responses of thrips (*Scolothrips takahashii*) foraging on hawthorn spider mites (*Tetranychus viennensis*), with data from Ding-Xu et al. (2007) as presented in FoRAGE. Functional responses overall increased with temperature for both females (**A**) and males (**B**). However, the space clearance rate of thrips increased and then decreased with temperature (**C**), while handling time declined with temperature (**D**).

females than males (30 vs. 25°C), suggesting variable effects of temperature on the components of the foraging process. More recently, using the large compilation of functional responses from the FoRAGE database, and after accounting for body size of both predator and prey as well as taxonomic identity, space clearance rates peaked at intermediate temperatures (Uiterwaal and DeLong, 2020). Somewhat surprisingly, the peak temperature was not very high (in the 20s °C), suggesting that even relatively common warm temperatures can begin to lower space clearance rates for many organisms. The difference between a space clearance rate that continues to rise with temperature and one that starts to decline at an intermediate temperature has substantial implications for the impact of climate change on predator–prey interactions (DeLong and Lyon, 2020; Gilbert et al., 2014), but researchers have not tested many predator–prey pairs for the full dependence of the functional response on temperature.

3.4 Breaking down the handling time

The components of handling time come mostly from killing (t_k), ingesting (t_e), and digesting (t_d) food. The handling time also generally must include any loss of search time arising from unsuccessful attacks (t_u), as these attacks take time away from searching for other prey. In some cases, handling time might include transporting food to a mate, nest, or nestling, if the predator is a central place forager, as well as recovery time (t_r) that might delay the predator in getting around to searching again. Thus, we can break down the handling time minimally as:

$$h = t_k + t_e + t_d + t_u + t_r \qquad (3.3)$$

It should not be surprising, then, that the handling time estimated with a functional response is different than the handling time estimated by observing the time it takes for a predator to "handle" its prey (Tully et al., 2005). Handling prey is, confusingly, only part of "handling time." The time during which one can see a predator handling prey might be better referred to as "detection" time, which turns out to be useful for estimating functional responses in the field (see Chapter 11 and Novak et al., 2017), and we might generally expect the detection time to be smaller than the handling time estimated by fitting a functional response curve. A somewhat open question is how one deals with the time that a predator might need to allocate to non-hunting behaviors, such as mating, tending to young, or hibernating. Relatively short-term foraging experiments will typically capture handling time without these other time costs that might cut into searching time (Li et al., 2018). This arises from the fact that most foraging experiments last only a small fraction of a predator's life. Estimating handling time from long-term time series (i.e., over multiple generations) in which predators have time to do other things may alter the estimates of handling times (see Chapter 11). Nonetheless, handling time still means the same thing—time that cuts into searching per prey captured.

Some of the more obvious traits that should be linked to handling time are prey body size, as a larger prey is likely to be more difficult to subdue and take longer to consume and digest than a smaller one (Kalinoski and DeLong, 2016; Samu, 1993). Similarly, larger predators may have an easier time subduing prey or have a larger digestive capacity. Not surprisingly, then, it is often the case that handling time decreases with predator body size across predator–prey pairs, again following a power-law-like relationship (DeLong et al., 2015; Miller et al., 1992; Rall et al., 2012). The slope of the scaling relationship between predator and prey body mass and handling time is highly variable (Uiterwaal and DeLong, 2020). In some cases, handling time might not even change at all with predator body size, as seems to be the case for mammalian carnivores, possibly because larger mammalian predators take increasingly larger prey, canceling out the handling time benefit of being larger (Carbone et al., 2007; DeLong and Vasseur, 2012a). These body size effects also show up within predator–prey pairs. For example, damselfly larvae (*Ischnura elegans*) foraging on *D. magna* had handling times that increased with prey size and decreased with predator size (Thompson, 1975). This effect may have been linked to the tight relationship between the size of the predator's gape and that of their overall body size, because the gape presumably sets how easy it is to consume prey of a particular size. Similarly, the predatory mite *Neoseiulus womersleyi* had longer handling times when foraging on later instars of spider mites (*Tetranychus macfarlanei*) (Ali et al., 2011).

Prey defensive traits such as dangerous morphologies (e.g., spines), avoidance behaviors, or toxins also should increase handling times (Hammill et al., 2010; Kalinoski and DeLong, 2016). For example, the neckteeth of *Daphnia* appeared to reduce the space clearance rate of *Chaoborus* predators but also increased the handling time when they did successfully kill *Daphnia*, judging by the lowered asymptote of their functional response on the necktooth morph individuals (Jeschke and Tollrian, 2000). Similarly, copepod predators tended to have longer handling times when foraging on prey with exoskeletons (e.g., insects and crustaceans) than when foraging on prey without exoskeletons (e.g., protists and worms) (Kalinoski and DeLong, 2016).

Although hunger level (predator satiation) appears to be a factor in motivating predators to forage, it also seems to reduce the time a predator spends handling its prey. Li et al. (2018) showed across numerous functional responses that longer food deprivation leads to steeper functional responses but shorter handling times. This finding is somewhat counterintuitive, as the amount of energy or nutrients extracted from an individual prey item should increase the longer a predator spends processing it. One might imagine that a hungry predator would attempt to maximize the energy gain from a prey capture, but an alternative hypothesis is that hungry predators faced with an abundance of food just consume the most nutritious parts of their prey to allow time for more captures.

As with space clearance rate, handling time may vary with temperature. The typical pattern is that handling times decline with temperature, as digestion rates or other biomechanical steps involved in processing prey may proceed more quickly at higher temperatures, reducing the time it takes to perform those functions (Figure 3.6D). This decline in handling time with temperature has been found across predator–prey pairs in meta-analyses (Kalinoski and DeLong, 2016; Rall et al., 2012; Uiterwaal and DeLong, 2018). The decline in handling time with temperature also has been found within species for spotted ladybird beetles (*Coleomegilla maculata*) feeding on green peach aphid (*Myzus persicae*) (Sentis et al., 2012), dragonflies (*Celithemis fasciata*) feeding on midges (*Chironomus tentans*) (Gresens et al., 1982), the bug *Cyrtorhinus lividipennis* foraging on the brown planthopper (*Nilaparvata lugens*) (Song and Heong, 1997), the spider *Anyphaena accentuata* and a variety of crab spiders (*Philodromus* spp.) foraging on pear psylla (*Cacopsylla pyri*) (Pekár et al., 2015), and for perch (*Perca fluviatilis*) and roach (*Rutilus rutilus*) foraging on phantom midges (*Chaoborus obscuripes*) (Persson, 1986). Using the FoRAGE database, however, across a very large set of functional responses, a u-shaped relationship between handling time and temperature emerged, indicating that handling times are minimized at intermediate environmental temperatures (Uiterwaal and DeLong, 2020). Whether handling time generally would start to increase with hotter temperatures within predator–prey pairs is not clear at this time, but this pattern was seen for handling time of the marine flagellate *Oxyrrhis marina* foraging on phytoplankton (*Isochrysis galbana*), indicated by a peak in the asymptotic foraging rate at intermediate temperatures (Kimmance et al., 2006).

3.5 Limits on the parameter space

There are many ways in which limits could arise on the values that functional response parameters may take. The first clear boundary is that both space clearance rate and handling time must be positive numbers, so the more interesting boundaries are the upper ones. How steep or high can a functional response be? What might constrain them to be lower? One clear source of limits is mechanical. There may be many prey types that predators are physically incapable of capturing or processing, putting a limit on space clearance rate through the probability of success (or the decision to not attack) (Portalier et al., 2019). Similarly, predators require morphological or physiological structures to handle their prey when captured, and to the extent they are able to evolve these, they may be forced to spend a certain amount of time handling prey, generating a lower bound on handling time.

Alternatively, because space clearance rate and handling time determine the strength of interaction between predator and prey, it is possible that there are combinations of parameters that either (1) do not allow predators to acquire enough energy to persist or (2) allow them to overexploit their prey to extinction. Either of these outcomes would tend to lead to extinction of predator populations, prey populations, or both, leaving only a subset of predator–prey pairs persisting. If this subset represents certain combinations of space clearance rates and handling times, then what we would see in nature is more restricted than what is mechanically feasible. This process of trimming unstable ecological interactions from a food web is known as stability selection (Borrelli et al., 2015), and it has been invoked to explain negative correlations among space clearance rates and handling times (Johnson et al., 2015), the persistence of communities (Guzman and Srivastava, 2019), and the scaling of predator body mass with prey body mass (DeLong, 2020; Guzman and Srivastava, 2019; Portalier et al., 2019; Weitz and Levin, 2006).

Looking at all the known space clearance rates and handling times from predator–prey pairs from around the world housed in the FoRAGE database (Figure 3.1), it appears that there may be some natural boundaries on the values they can take, but a strong negative correlation across all parameter values is not apparent. Controlling for more factors, however, indicates that just such a negative correlation between space clearance rate and handling time does occur (Kiørboe and Thomas, 2020). Since the parameters emerge from the interaction among predator and prey, this correlation cannot be ascribed simply to a continuum of predator strategies, as suggested in Kiørboe and Thomas (2020). Rather, it reflects the influence of mechanical and physiological constraints, abiotic factors, predator and prey traits, and any strategies that both predator and prey implement in the foraging process. Furthermore, we are probably not ready to think of the observations in Figure 3.1 as fully representative of nature, because it is likely that this distribution is strongly influenced by the propensity of researchers to measure the functional responses of predator–prey pairs that are big enough to be seen with the naked eye but small enough to forage in desktop or other laboratory arenas without habitat complexity. The actual relationship between space clearance rate and handling time may not be seen until functional response measurements include more types of organisms or achieve coverage of most of the links in specific food webs.

3.6 Other predators

As indicated in Chapter 2, functional responses may decline with predator abundance, and how this happens is a major open question. However, for many years, *whether* additional predators reduce the functional response was a major debate. The major sides of the debate were that functional responses are prey dependent (meaning only dependent on prey abundance, as in the standard Holling disc equation) or that functional responses are ratio dependent. Ratio dependent means that the functional response was not written as just a function of R but of R/C (equivalent to a mutual interference level of -1). There has been considerable back and forth on this issue (Abrams, 1994, 2015; Abrams and Ginzburg, 2000; Akcakaya et al., 1995; Ginzburg and Jensen, 2008), and I will not rehash this debate here. There are, however, two important points about this debate relevant to this book. First, most of the studies that have examined whether functional responses decline with predator density have found that they do (DeLong and Vasseur, 2011; Skalski and Gilliam, 2001). In most cases, the functional response is neither exactly ratio dependent nor only prey dependent, but predator dependent, which just means that the functional response does decline with predator density in some way. Second, why there is predator dependence has not been worked out. The two most commonly used modifications of the functional response to include predator density make different hypotheses. The Hassell–Varley model invokes mutual interference, wherein the space clearance rate is a power-law function of predator density (Hassell and Varley, 1969). As a result, in this view, predator density could be altering velocity, detection distance, and the probabilities of attack and success. None of these connections has been made empirically. One study found that across replicate populations of the protist *Didinium nasutum* consuming the protist *Paramecium aurelia*, space clearance rate was positively correlated with the magnitude of interference, and that a dependence of search velocity on predator density could account for this pattern (DeLong and Vasseur, 2013). The Beddington–DeAngelis model, on the other hand, hypothesizes that other predators cut into searching time with an amount of time wasted per predator in the population (Beddington, 1975; DeAngelis et al., 1975). Wasted time could be as simple as having to reroute a search path to physically go around a conspecific. Thus, some searching time would be wasted if the predator cannot find food while rerouting. Thus, this model invokes a change to the time budget, but predators continue to behave in the same way in the presence of other predators when they do search. In addition to the variety of potential mechanisms that generate interference, there is little work evaluating sources of variation in the magnitude of interference. But once again, body size could play a role (DeLong, 2014). For example, for the mirid predator *Macrolophus pygmaeus* preying on flour moth (*Ephestia kuehniella*) eggs, interference was larger for larger predators (Papanikolaou et al., 2021).

So far, we have attributed variation in functional response parameters to aspects of a focal predator–prey interaction. Predators, however, also may be prey, and the risk of being killed while foraging may be a major factor driving predator functional responses. It is clear that risk is a major factor in foraging in general (Lima and Dill, 1990), but the extent to which this can be detected in a functional response is less clear. For example, larval spotted salamanders (*Ambystoma maculatum*) from ponds with high risk of predation from marbled salamanders (*Ambystoma opacum*) showed higher space clearance rates than spotted salamanders from low-risk ponds (Urban et al., 2020). Functional responses from high-risk ponds also were more type III than those from low-risk ponds. Given that prey at any trophic level may adjust their behavior to reduce risk, there is a clear need for further assessment of how functional responses are altered by the presence of other predators (Cuthbert et al., 2020a; Toscano and Griffen, 2014). As above, this implies changes in the space clearance rate or handling times relative to the scenario when predators are alone.

Such differences are sometimes termed multi-predator effects (MPEs), and this includes risk reduction (prey are safer with more predator types around) or risk enhancement (prey are at greater risk with more predators around) (McCoy et al., 2012; Sih et al., 1998). One clear way in which this could happen is that a predator who is itself at risk of predation could spend less time searching and thus reducing risk, but then experiencing a lower encounter rate with its own prey, reducing food intake. Alternatively, prey that find themselves avoiding some predators may inadvertently become more susceptible to others, through changes in their detectability (d) or ability to employ defensive behaviors or occupy safer habitats (Miller et al., 2014). In either case, shifts in functional responses have not been well studied with respect to predation risk to the forager.

4

Population Dynamics and the Functional Response

In this chapter, I show how the functional response can drive predator–prey cycles (and dynamics more generally). I introduce predator–prey differential equation models and fit them to real dynamic data from classic predator–prey systems (lynx–hare and *Daphnia*–algae). This coupling achieves two things. First, it allows me to demonstrate that the models are capable of describing real predator–prey dynamics. Second, it allows me, from an empirically grounded starting point, to vary the parameters of the functional response to show how changes in the functional response parameters change the dynamics.

4.1 The functional response as a trophic link

The functional response is the link connecting trophic levels in food chains and food webs. It determines the rate of foraging at any given density of the prey and predator and therefore tracks the flow of energy from the prey population to the predator population through consumption. Furthermore, the functional response simultaneously describes how much the predators can reduce the abundance of the prey population and how the consumption of prey fuels the production of more predators (Urban et al., 2020). Although there remains some controversy about what causes predator–prey cycles in nature, it is clear that the functional response drives decreases and increases in abundance that can drive cycles in predator–prey systems (Krebs et al., 2001; Turchin, 2001).

One common way of describing the increases and decreases in both predator (C) and prey abundance (R) through time is with coupled ordinary differential equation models (ODEs). The simplest of these models is the classic Lotka–Volterra (LV) predator–prey model (Hastings, 1997) (Figure 4.1). In this model, the prey birth rate is set by exponential growth (rR), that is, population growth with no inherent density dependence, where r is

Predator Ecology: Evolutionary Ecology of the Functional Response. John P. DeLong, Oxford University Press.

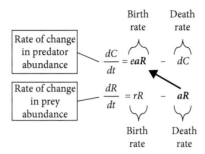

Figure 4.1 Schematic of the Lotka–Volterra predator–prey model. This model and others like it couple two equations that describe the change in the abundance of prey and predator through time. Each term in the model is either the birth rate or death rate of that species given some biological process. In a predator–prey model, the equations are connected by the functional response (in this case, a type I, bolded), which simultaneously drives the prey population down and fuels the birth rate of predators.

the intrinsic rate of population growth. The functional response is type I (linear without a ceiling), and it simultaneously describes a reduction in prey abundance (its effect in the prey equation is negative) and an increase in predator abundance by fueling the births of predators through the conversion efficiency e (i.e., the functional response links the two populations). The conversion efficiency tells you how many fully functioning adult predators are produced per prey consumed;[1] it is usually a very small number and always less than one as long as the prey and predator are expressed as whole organisms and not biomass. Finally, the predators die at some constant rate d.

4.2 Adding some complexity

The LV model includes no mechanism for self-limitation in the prey or the predator population, so the LV is widely viewed as unrealistic. Surely, without predators, most prey would eventually reach some maximum or relatively stable average number (i.e., a carrying capacity). We can add a generic kind of environmental limitation by changing the prey growth term from an

---

[1] Yes, it is true that organisms are not generally born as fully functioning adults. Nonetheless, this is the assumption of all predator–prey models that do not have separate equations for juveniles and adults (those are stage-structured models). This simplifying assumption still allows us to characterize dynamics of some populations reasonably well because in these cases the conversion efficiency implicitly reflects the energetic cost of growth and maturation.

exponential model to the logistic growth model $\left(rR\left(1 - \frac{R}{K}\right)\right)$, where r is the same and K is the prey carrying capacity:

$$\frac{dC}{dt} = eaRC - dC \qquad (4.1)$$

$$\frac{dR}{dt} = rR\left(1 - \frac{R}{K}\right) - aRC \qquad (4.2)$$

This model is already complex enough to describe the concurrent dynamics of two-species predator–prey interactions such as those of water fleas (*Daphnia ambigua*) eating algae (*Scenedesmus obliquus*) in laboratory conditions. To illustrate, I fit equations (4.1) and (4.2) to data on the abundances of *Daphnia* and algae from a mesocosm experiment (DeLong et al., 2018) to determine how well the model could describe the dynamics of a real system (or at least a laboratory microcosm housing real organisms) (Figure 4.2). Fitting an ODE to a time series is an iterative process that solves the equations for a wide range of parameter values and finds the set of solutions (time series of abundances) that most closely aligns with the data (here I used the Potterswheel toolbox; described in Maiwald and Timmer, 2008). When a set of parameters can be found such that the solution to the model looks like the data, this suggests that the processes captured in the model (e.g., the functional response and

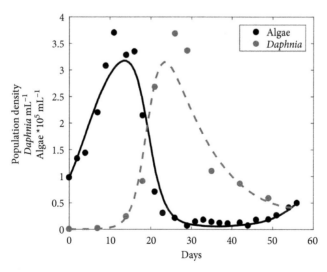

Figure 4.2 Population dynamics of a *Daphnia*–algae cycle, using data from DeLong et al. (2018). Densities of *Daphnia* and algae along with fits of equations (4.1) and (4.2) to the data.

prey reproduction) are at least a good hypothesis for what is actually driving the predator–prey cycles.

Switching the functional response in equations (4.1) and (4.2) from the type I to the type II $\left(\frac{aR}{1+ahR}\right)$ results in the MacArthur–Rosenzweig model (MR; Rosenzweig and MacArthur, 1963):

$$\frac{dC}{dt} = e\frac{aRC}{1 + ahR} - dC \tag{4.3}$$

$$\frac{dR}{dt} = rR\left(1 - \frac{R}{K}\right) - \frac{aRC}{1 + ahR} \tag{4.4}$$

Although the MR model is a common starting point for theoretical ecology, let's make the model a little bit more complicated by adding mutual interference; that is, when the average per capita foraging rate declines with more predators. Remember from Chapter 2 that one way of adding interference is as simple as making the space clearance rate a function of predator density (recall equation (2.11)). Updating our model, where m again is mutual interference, we get:

$$\frac{dC}{dt} = e\frac{aRC^{m+1}}{1 + ahRC^m} - dC \tag{4.5}$$

$$\frac{dR}{dt} = rR\left(1 - \frac{R}{K}\right) - \frac{aRC^{m+1}}{1 + ahRC^m} \tag{4.6}$$

Adding interference to the model reduces the height of the functional response as predators increase in abundance. This decrease then dampens down the cycles because foraging rates slow as the predator population increases, leaving some prey unharvested. These kind of dampened cycles are apparent in the dynamics of lynx (*Lynx lynx*) and snowshoe hare (*Lepus americanus*) reported in the study by Slough and Mowat (1996) (Figure 4.3). In this study, the typical oscillating pattern shown by predators and prey (at least for predators and prey that have a strong trophic interaction) is apparent for about one full cycle. Again, I used ODE fitting to find parameters of equations (4.5) and (4.6) that generate a match between the model and the data, and found that the model fits pretty well, again suggesting that the functional response is a key part of the process causing lynx and hare populations to cycle (Krebs et al., 2001). The cycle plays out when hares grow more numerous when lynx are rare, allowing lynx to slowly increase as well. Eventually, the hare population starts to decline and the lynx population crashes after them, as lynx numbers are no longer replaced when hare populations are depressed and consumption is reduced. When the lynx and

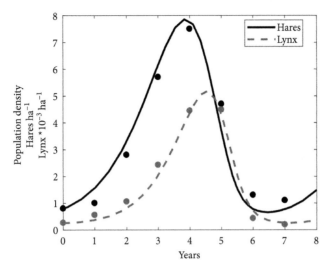

Figure 4.3 Population dynamics of a lynx–hare cycle, using data from Slough and Mowat (1996). Densities of lynx and hare and fits of equations (4.5) and (4.6) to the data.

hare are again rare, there is little prey mortality and close to exponential growth, sending the hare population back up into another round of the cycle.

The role of the functional response in these dynamics is to determine how fast predators can reduce the population size of the prey and how prey consumption leads to more predators. Thus, a steeper or higher functional response can alter the amplitude of cycles. To illustrate, starting with the fitted parameters for the lynx–hare system, I varied the space clearance rate and handling time just 1% above and below the fitted value, and the cycles changed shape (Figure 4.4B,C). The effect of variation in the space clearance rate on the amplitude of the cycle was relatively slight, but by comparison, varying the handling time had a relatively large effect (Figure 4.4D,E). In general, the magnitude of the effect a change in a parameter has on the dynamics may depend on the starting value of the parameter and whether the cycles already allow high prey density, which is when handling time becomes an increasingly important constraint. Increasing the space clearance rate lowered the height of the cycle's peaks for both predator and prey, while increasing the handling time increased the height for both. This counterintuitive result arises because more foraging (owing to higher space clearance rate or lower handling time) limits the growth of the prey, reducing the size of their peak and thus the amount of prey the predator can get out of the system over the long run, reducing the height of the predator peak as well.

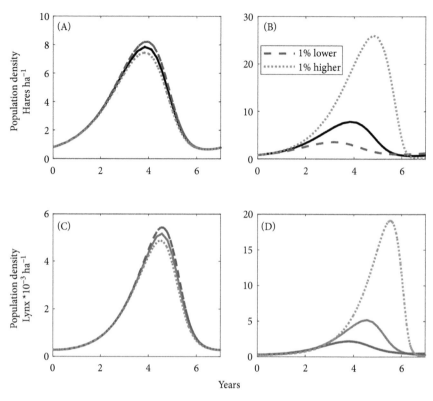

Figure 4.4 Population dynamics of a lynx–hare cycle, using data from Slough and Mowat (1996). Each panel shows fits of equations (4.5) and (4.6) to the lynx–hare data shown in Figure 4.3. Also shown are solutions where the parameters are 1% less than (dashed lines) and 1% more than (dotted lines) estimated from fitting. Space clearance rate is varied in **A** and **C**; handling time is varied in **B** and **D**. Hare abundances are in **A** and **B**; lynx abundances are in **C** and **D**. Note that the y axes have different units in the different panels.

Figures 4.2 and 4.3 suggest that predator and prey populations may not just cycle randomly but rather they may revolve around a particular value like planets around the sun (Ginzburg and Colyvan, 2004). This center of gravity is an "equilibrium" point and often occurs roughly at the average density. Of course, changing the parameters in the model changes the point around which the populations cycle and the shape of those cycles. Analyzing the equilibria of models is a useful way of understanding predator–prey cycles. The approaches used for this are well-described elsewhere (Hastings, 1997; McCann, 2011; Otto and Day, 2007), so here I just point out that when you see cycles, you should pay attention to how they cycle and what they cycle around. The functional response plays an important role in determining the shape of those orbits around the equilibrium as well as the location of the equilibrium. In our two examples, *Daphnia* consuming algae and lynx consuming hares, the cycles have very different shapes. In Figure 4.5, the cycles are plotted in a phase-plane with the prey density on the x axis and the predator density on the y axis, overlaid by a vector field showing the direction the system (predator and prey abundances together) would move if you started the system at any point in the phase-plane (i.e., any combination of predator and prey abundance). A key difference between the two models that describe

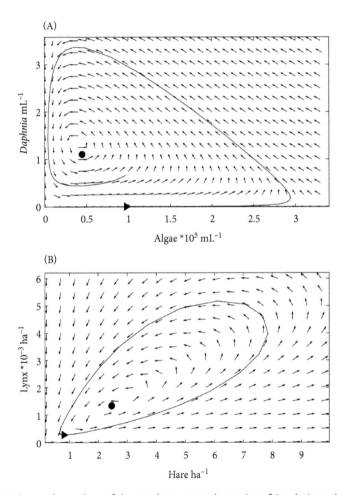

Figure 4.5 Phase-plane plots of the predator–prey dynamics of *Daphnia* eating algae (**A**) and lynx eating hare (**B**). The filled circle in the middle is the equilibrium, and the filled triangle is the starting point of the time series. The arrows represent the direction of movement for both populations together at any particular combination of abundances.

these predator–prey dynamics is the functional response, with a type I in the *Daphnia*–algae system (Figure 4.5A) and a type II with interference in the lynx–hare system (Figure 4.5B). (They also differ in the prey growth term). In the *Daphnia*–algae system, the type I functional response allows the *Daphnia* to rapidly crash out the algae as the *Daphnia* grow (the big vertical drop in Figure 4.5A). The presence of a handling time and interference competition, however, dramatically dampens foraging rates and allows the lynx and hare to grow together for some time before the cycle peaks and declines (Figure 4.5B). Thus, the form, shape, and parameters of the functional response have implications for the abundance and dynamics of predator and prey populations, and by extension, whole communities (Fussmann and Blasius, 2005).

5
Multi-species Functional Responses

In this chapter, I extend the standard functional response model to communities in which predators are foraging on more than one kind of prey. This is an essential component of real foraging scenarios that is not yet widely represented in the functional response literature. Here I develop the multi-species functional response (MSFR) and describe it using my particular perspective on how we understand these functions in general.

5.1 The need for MSFRs

Functional responses that describe the foraging of one type of predator on one type of prey have been studied a great deal. Most predators, however, forage on at least a few prey types, if not many types. For example, aplomado falcons (*Falco femoralis*) prey on over 45 species of birds plus bats and some insects (Hector, 1985). Likewise, the wolf spider *Pardosa glacialis* selects prey from at least 51 different arthropod families (Eitzinger et al., 2019). Specialist predators that forage on only one or a few prey types are out there, such as the ciliate predator *Didinium nasutum* that specializes on *Paramecium* spp. and snail kites (*Rostrhamus sociabilis*) that specialize on snails, but they appear to be rather rare. Furthermore, some specialists may have a broad diet for like prey (García et al., 2018), suggesting that specialization may not occur at the prey species level but among prey types that are similar in the way the predator perceives them (i.e., *Didinium* is a *Paramecium* specialist but eats any type of *Paramecium*). Although specialist predators may play a role in driving classic predator–prey cycles (Hanski et al., 1991; Oli, 2003), they may not be numerically important with respect to the movement of energy through most food webs.

Generalist predators are thought to have very different effects on the stability and dynamics of ecological communities than specialist predators (Hanski et al., 1991), and foraging on multiple prey can have important consequences for community diversity and interactions among prey species

Predator Ecology: Evolutionary Ecology of the Functional Response. John P. DeLong, Oxford University Press.
© John P. DeLong 2021. DOI: 10.1093/oso/9780192895509.003.0005

(Downing, 1981; Johnke et al., 2017; Wahlström et al., 2000). For example, golden eagles (*Aquila chrysaetos*) on the Channel Islands may forage on island spotted skunks (*Spilogale gracilis amphiala*), island foxes (*Urocyon littoralis*), and introduced feral pigs (*Sus scrofa*). This multi-way interaction has implications for the persistence of golden eagles on the islands and their predatory impacts on the foxes (Roemer et al., 2002). Because foraging on feral pigs helps to bolster the golden eagle populations, there is heightened predation on foxes, whose populations are small and of conservation concern. Similarly, the ability of the wolf spider *Tasmanicosa leuckartii* to function as a biocontrol agent on *Helicoverpa armigera* larvae in cotton fields depends on the larvae's interactions with other prey such as *Hogna crispipes* (another wolf spider and a potential intraguild predator, which is a predator that eats both a predator and that predator's prey) and the ground cricket *Teleogryllus commodus* (Rendon et al., 2019). Similar complexities arise for acarine predators foraging on alternative prey in a biocontrol scenario for apple pests (Lester and Harmsen, 2002). Furthermore, classic ecological processes such as keystone predation depend on predators being able to consume a wide range of prey types and as a result can generally consume the most numerically dominant prey the most (Paine, 1966). In some cases, the totality of all these interactions is difficult to predict by simply aggregating across the pairwise interactions, suggesting that predators change their rate of foraging on some prey in the presence of other prey (Colton, 1987).

The ability to forage on a wide range of prey types has implications for predators as well. In bromeliad tank ecosystems, for example, predatory damselfly larvae (*Leptagrion andromache*) showed greater population persistence when prey types were more diverse (Guzman and Srivastava, 2019). Furthermore, being able to access a diverse prey base can allow predators to maintain energy uptake that supports reproduction and survival, as in common guillemots (*Uria aalge*) that select among a variety of forage fish that vary in abundance, and in so doing, maintain sufficient energy delivery to their chicks (Smout et al., 2013). Bonelli's eagles (*Aquila fasciata*) had greater reproductive success and survival when foraging mostly on their main prey, European rabbits (*Oryctolagus cuniculus*), but when these were rare, they turned to alternative prey even though reproduction was reduced (Resano-Mayor et al., 2016). Some predators may select among prey types to balance specific required nutrients, with implications for their fitness. Invertebrate predators such as ground beetles and spiders, for example, may preferentially pick prey types to balance their lipid and protein intake (Mayntz et al., 2005; Schmidt et al., 2012), and this nutrient balancing can be critical to reproductive success (Wilder and Rypstra, 2008).

It is therefore essential to expand work on functional responses to include multiple prey types (Chan et al., 2017; Kalinkat et al., 2011). Such MSFRs can be difficult to estimate (Gentleman et al., 2003). Indeed, there are few generalist predator species for which even multiple single-species functional responses have been estimated. The predators for which functional responses have been measured on many prey types tend to be economically important, such as some widely used biocontrol arthropods like the ladybird beetle *Harmonia axyridis*, which also happens to be invasive and impacting native ladybird beetles (Aqueel and Leather, 2012; Lee and Kang, 2004; Uiterwaal and DeLong, 2018). Studies also frequently focus on predators that are relatively easy to work with, such as the widespread predatory marine flagellate *Oxyrhis marina* (Roberts et al., 2010). In one notable case, however, a large accumulation of laboratory-based pairwise functional responses for marine protists consuming a range of algae and other protists was used as the basis of constructing a planktonic food web (Jeong et al., 2010). Despite this, only a few actual MSFRs have been estimated (Baudrot et al., 2016; Colton, 1987; Hellström et al., 2014; Novak et al., 2017; Smout et al., 2010, 2013; Smout and Lindstrm, 2007), and these efforts clearly illustrate the empirical challenge, especially in field contexts, even though they generally are still focusing on a very small subset of the potential prey (often only two or three prey types). Even in some cases where MSFRs could be estimated, it may be analytically easier to break them into separate pairwise functional responses that are simply dependent on the density of more than one prey type (Chan et al., 2017). One particular challenge is that the abundance of prey is often only known in field settings as an index that lacks a clear spatial dimension such as prey per km^2 (Baudrot et al., 2016; Quinn et al., 2003; Smout et al., 2010). Even without much empirical estimation, MSFRs are widely invoked in food web models, indicating a critical need for improvement in this area (Berlow et al., 2009; Brose et al., 2006; Gentleman et al., 2003; Guzman and Srivastava, 2019; Morozov and Petrovskii, 2013; Petchey et al., 2008). MSFR model parameterization often relies on simplifying assumptions such as proportion-based or optimally foraging predators (Morozov and Petrovskii, 2013), but there is little empirical support for any of these modeling choices from empirically estimated MSFRs.

5.2 Extending the functional response to multiple prey types

The MSFR is harder to both visualize and estimate than the single-species functional responses, since the function has both multiple dependent variables (foraging rate on different prey types) and independent variables (abundance of different prey types). But the key elements are the same in the pairwise and MSFRs. Indeed, we already have worked through the components of the standard MSFR, and all we need to do is sum up the foraging across multiple prey types and account for the handling time devoted to each prey type. Going back to our original derivation of the type II functional response, we had divided the total time into time spent handling and time spent searching:

$$T_{tot} = T_s + aRT_sh \tag{2.3}$$

If a predator is foraging on two types of prey, we need only add an additional amount of handling time for kills of the second prey type. To do this, we now modify parameters and abundances with a subscript for prey type 1 or 2:

$$T_{tot} = T_s + a_1R_1T_sh_1 + a_2R_2T_sh_2 \tag{5.1}$$

Following the same steps as before, the MSFR could be expressed as two equations, one for each prey type:

$$f_{pc,1} = \frac{a_1R_1}{1 + a_1R_1h_1 + a_2R_2h_2} \tag{5.2}$$

$$f_{pc,2} = \frac{a_2R_2}{1 + a_1R_1h_1 + a_2R_2h_2} \tag{5.3}$$

Or to expand it to any number of j prey types, the foraging rate on prey type i is:

$$f_{pc,i} = \frac{a_iR_i}{1 + \sum a_jR_jh_j} \tag{5.4}$$

The main thing to notice is that when focused on the consumption of a particular prey type, we have to account for the time spent handling all prey types. The total foraging rate on all prey types sums up equation (5.4) across the prey types and so becomes:

$$f_{\text{pc}} = \frac{\sum a_j R_j}{1 + \sum a_j R_j h_j} \tag{5.5}$$

This MSFR equation can be used to get a handle on what predators do when they have multiple prey types to consume and how this drives predator and prey population dynamics in a community rather than in just a pair of interacting species (Petchey et al., 2008; Smout and Lindstrm, 2007).

For multi-prey experiments in which prey depletion occurs, the MSFR would then be the multi-species version of the Rogers Random Predator (RRP) equation (Murdoch, 1969). The two-species version is:

$$F_{\text{pc},1} = R_{\text{o},1}\left(1 - e^{a_1\left(h_2 F_{\text{pc},2} + h_1 F_{\text{pc},1} - T_{\text{tot}}\right)}\right) \tag{5.6a}$$

$$F_{\text{pc},2} = R_{\text{o},2}\left(1 - e^{a_2\left(h_2 F_{\text{pc},2} + h_1 F_{\text{pc},1} - T_{\text{tot}}\right)}\right) \tag{5.6b}$$

These equations of course are not closed-form solutions for the number of either prey type, requiring the use of numerical solvers to fit and/or predict the number of prey eaten. To check whether the Lambert Random Predator (LRP) equation can be extended to two species and thereby simplify matters, I extended Bolker's (2011) derivation of the original LRP to two prey types in Box 5.1. We can indeed write out a similar expression this way, but the resulting form of the model still requires numerical solvers to get the abundance of the two types across the two equations, so it is not quite as useful as the single-prey version.

Box 5.1. Derivation of the two-species Lambert$W$ random predator equation.

Going back to Bolker (2011) for our starting point, the two-species RRP for the number of prey species 1 eaten is the following, where parameters are subscripted 1 or 2 for prey types 1 and 2, respectively:

$$F_{\text{pc},1} = R_{\text{o},1}\left(1 - e^{a_1\left(h_2 F_{\text{pc},2} + h_1 F_{\text{pc},1} - t\right)}\right)$$

Rearranging we can get:

$$1 - \frac{F_{\text{pc},1}}{R_{\text{o},1}} = e^{-a_1\left(t - h_2 F_{\text{pc},2} + h_1 F_{\text{pc},1}\right)}$$

Multiply out the exponent:

$$1 - \frac{F_{pc,1}}{R_o} = e^{-a_1 t + a_1 h_2 F_{pc,2} + a_1 h_1 F_{pc,1}} = e^{-a_1 t} e^{a_1 h_1 F_{pc,1}} e^{a_1 h_2 F_{pc,2}}$$

As with the single-prey RRP, multiply the right-hand side by $\frac{e^{a_1 h_1 R_{o,1}}}{e^{a_1 h_1 R_{o,1}}}$, combine the first exponential term with the numerator and the second with the denominator, and leave the last exponential term alone:

$$1 - \frac{F_{pc,1}}{R_o} = \left(e^{-a_1 t} e^{a_1 h_1 R_{o,1}}\right) \left(\frac{e^{a_1 h_1 F_{pc,1}}}{e^{a_1 h_1 R_{o,1}}}\right) e^{a_1 h_2 F_{pc,2}}$$

Now follow exponential rules and factor out:

$$1 - \frac{F_{pc,1}}{R_o} = \left(e^{-a_1(t - h_1 R_{o,1})}\right) \left(e^{a_1 h_1 (F_{pc,1} - R_{o,1})}\right) e^{a_1 h_2 F_{pc,2}}$$

As before, multiply the exponent in the second exponential term by $\frac{R_{o,1}}{R_{o,1}}$, which allows rearranging to:

$$1 - \frac{F_{pc,1}}{R_o} = \left(e^{-a_1(t - h_1 R_{o,1})}\right) \left(e^{a_1 h_1 R_{o,1}\left(\frac{F_{pc,1}}{R_{o,1}} - 1\right)}\right) e^{a_1 h_2 F_{pc,2}}$$

Finally, multiply both sides by $a_1 h_1 R_{o,1}$ and move the second exponential term to the left-hand side:

$$\left[a_1 h_1 R_{o,1}\left(1 - \frac{F_{pc,1}}{R_{o,1}}\right)\right] e^{\left[a_1 h_1 R_{o,1}\left(1 - \frac{F_{pc,1}}{R_{o,1}}\right)\right]} = a_1 h_1 R_{o,1}\left(e^{-a_1(t - h_1 R_{o,1})}\right) e^{a_1 h_2 F_{pc,2}}$$

Again we see that this expression takes the form of the product log, which is that for some function $y = x e^x$, the solution is that $W(y) = x$. This means that:

$$a_1 h_1 R_{o,1}\left(1 - \frac{F_{pc,1}}{R_{o,1}}\right) = W\left(a_1 h_1 R_{o,1}\left(e^{-a_1(t - h_1 R_{o,1})}\right) e^{a_1 h_2 F_{pc,2}}\right)$$

which rearranges to equation (6.11):

$$F_{pc,1} = R_{o,1} - \frac{W\left(a_1 h_1 R_{o,1}\left(e^{-a_1(t - h_1 R_{o,1})}\right) e^{a_1 h_2 F_{pc,2}}\right)}{a_1 h_1}$$

which is the LambertW form of the RRP for two species for prey type 1. As with the two-species RRP, however, there are still two unknowns, and so there is not yet a closed-form version like the single-species LRP.

MSFRs have been measured empirically in only a few instances, but each of these instances has revealed something interesting about predation in a multi-species context. A field-based MSFR for hen harriers (*Circus cyaneus*) foraging on birds and mammals was instrumental in revealing how the availability of alternative prey was important for the management of red grouse (*Lagopus lagopus*), a game bird upon which harriers were foraging and which was a focal target of local hunters (Smout et al., 2010). In particular, if alternative prey are available (pipits and voles), harriers focus less on grouse, allowing grouse populations to remain larger, facilitating continued harvest by humans. Using a similar approach, Smout et al. (2013) showed that the MSFR of common guillemots (*Uria aalge*) foraging on small fishes was such that the guillemots were able to gather sufficient prey to feed chicks by catching whatever fish species were available. Minke whales (*Balaenoptera acutorostrata*), although foraging on fish and crustaceans, appear more likely to lower the abundance of capelin (*Mallotus villosus*) than other species, given the MSFR (Smout and Lindstrm, 2007). Finally, barn owls (*Tyto alba*) and red foxes (*Vulpes vulpes*) both appear to increase consumption on some prey species when they are rare, leading to possible destabilization of the food webs (Baudrot et al., 2016). Undoubtedly, MSFRs are mediating many other interesting food web phenomena, but these remain undiscovered because MSFRs have not been well documented.

5.3 An example with damselfly naiads

A rare laboratory effort to estimate an MSFR was undertaken for naiads of the damselfly *Enallagma aspersum* foraging on two types of microcrustaceans, a cladoceran (*Simocephalus serrulatus*) and a copepod (*Diaptomus spatulocrenatus*) (Colton, 1987). This study suggested that high abundance of both prey types tended to allow damselflies to overconsume *Diaptomus* relative to *Simocephalus*, indicating a shift in the functional response of the damselflies on *Diaptomus* in the presence of *Simocephalus*.

To illustrate how this was done, I used data from this experiment and refit the single-species functional responses for each prey type using the LRP equation (equation (2.17); Figure 5.1A). From these data and fits, it is clear that damselfly naiads have higher foraging rates on *Simocephalus* than on *Diaptomus*. Then I plugged the parameters for those single-prey fits into the two-prey model (equations (5.6)) to predict foraging on both prey types given the other. Doing this implicitly assumes that damselflies do not change their foraging behavior when faced with multiple prey, making this

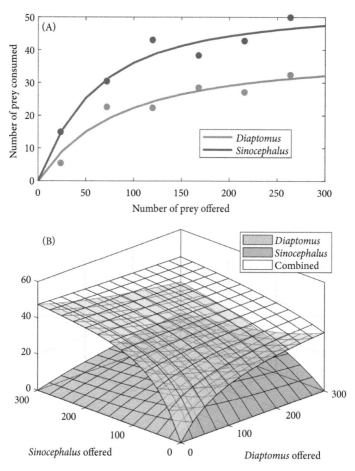

Figure 5.1 Single (**A**) and multi-species (**B**) functional responses for damselflies (*Enallagma aspersum*) foraging on copepods (*Diaptomus spatulocrenatus*) and cladocerans (*Simocephalus serrulatus*). Data from Colton (1987).

prediction a kind of null model for foraging on two prey types. The first thing to notice about the null-MSFR prediction is that the number of either prey type consumed goes down the more alternative prey was offered (shaded surfaces decline as alternate prey increases; Figure 5.1B). That is because, as parameterized, the predator will continue the same foraging behavior but pay handling time costs for consuming both prey types.

In the original study, Colton (1987) used this approach and then looked at actual foraging rates for damselflies given both prey types in a variety of combinations, with the null-MSFR as a baseline. The two-species foraging data showed that *Diaptomus* were consumed more than expected from the null-MSFR, but primarily when both prey types were abundant. This could be accounted for by an increase in space clearance rate for *Diaptomus* but a constant space clearance rate for *Simocephalus*. It is possible that damselflies showed some preference for *Diaptomus* (influencing p_a, see Chapter 8), but any component of space clearance rate for *Diaptomus* that increased with *Simocephalus* density could have caused this, from increased movements of *Diaptomus* (V_r) in a crowded foraging arena, to increased detection distance (d) or probability of success (p_s).

In this case, the total foraging expected to occur tops out very similar to the asymptote of *Simocephalus*, the prey species with the highest functional response. This pattern suggests a possible rule for a global handling time for generalist predators, which is that the total prey consumed cannot exceed $1/h$ of the prey type with the lowest handling time. This arises because at the asymptote of this prey type, most time is allocated to handling and almost none to searching. Adding other prey types would simply swap handling time for one species with that of another, longer-to-handle, prey type, lowering the search time and the total foraging rate. As a result, when adding prey with larger handling times, the asymptote has to go down below that of the prey type with the lowest h.

This particular experiment is a good example of the challenge of estimating MSFRs. In the study, six prey levels were used for each single-species functional response. To estimate the full MSFR, all 36 factorial combinations of prey levels were used. Thus, estimating one MSFR amounted to estimating the equivalent of six pairwise functional responses, and these were only two of the many potential prey types that could be consumed by the damselflies in an actual pond. Apart from the experimental effort, with prey depletion of both types of prey occurring during the experiment, fitting the MSFR to data required using the two-prey RRP equation. For single-species models, one could turn to the LRP equation, but with the two-prey model, a numerical solution is required to identify the two unknowns (number of each prey type) from the two equations simultaneously at all prey density combinations.

6

Selection on Functional Response Parameters

In this chapter I show why there should be selection on traits associated with functional response parameters. I describe this using standard quantitative genetics techniques to show how a classic evolutionary arms race arises and how it depends on key features of the functional response. I then show that selection on the predator and prey components of space clearance rate is synchronous for predator and prey through population cycles but alternating through time for handling time.

6.1 Why functional response parameters might change through evolution

Given the critical role space clearance rate and handling time play in predator–prey interactions and their direct effect on births and deaths (Jeschke et al., 2004; Urban et al., 2020), both parameters should be expected to be under strong selection. Critically, the functional response affects the overall change in population abundance for predator and prey as well as the per capita change in abundance. And the per capita population growth rate $\left(\text{e.g.,} \frac{1}{R}\frac{dR}{dt}\right)$ also happens to be a common metric for average fitness in a population, because it represents the average rate of individuals making copies of themselves or dying out (Abrams et al., 1993; Lande, 1982). Thus, the functional response is driving ecological dynamics at the same time as it is setting fitness for both the predator and prey, which is the foundational idea behind eco-evolutionary dynamics (Palkovacs and Hendry, 2010; Post and Palkovacs, 2009; Schoener, 2011). One would expect, then, that there would be natural selection for behaviors and morphologies that allow predators to have a higher functional response, which can be achieved by having a large space clearance rate and/or a low handling time. In contrast, one would expect there to be selection for prey to be good at avoiding predation, which

Predator Ecology: Evolutionary Ecology of the Functional Response. John P. DeLong, Oxford University Press.
© John P. DeLong 2021. DOI: 10.1093/oso/9780192895509.003.0006

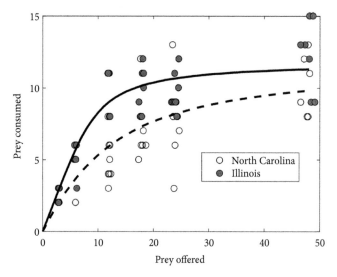

Figure 6.1 Functional response of tree hole mosquitos (*Toxorhynchites rutilus septentrionalis*) foraging on *Aedes triseriatus* mosquito larvae. Data from Livdahl (1979). The curve was lower for prey that were collected in the predator's geographic range, suggesting they may have evolved defenses against the predator, while the naïve prey were more susceptible.

means having a low functional response. This outcome can be achieved by having a small space clearance rate and, possibly, a high handling time.

Despite the expectation of selection altering functional responses, there is scant direct evidence for it. In spotted salamanders (*Ambystoma maculatum*) (Urban et al., 2020), functional responses were higher for individuals from ponds with higher risk of predation from marbled salamanders (*Ambystoma opacum*). This difference was detectable after rearing in a common garden environment, indicating a genetic basis for the difference. Similarly, the functional response of tree hole mosquitos (*Toxorhynchites rutilus*) foraging on *Aedes triseriatus* mosquito larvae differed according to the regional origin of the prey (Livdahl, 1979). The functional response was lower for prey that were collected from the tree hole mosquito's geographic range (North Carolina) than for prey outside of that range (Illinois) that thus had no previous exposure to these predators (Figure 6.1). This difference is consistent with the idea that the prey co-occurring with the predator had evolved traits that reduced their predation by this predator. However, Juliano and Williams (1985) cautioned against this interpretation, suggesting there may be other ecological ways in which such a difference could arise.

Remembering that space clearance rate and handling time reflect traits of both predator and prey (Abrams, 2000), selection should occur in both predator and prey through the effect of these parameters, forming the basis for co-evolution. This co-evolution is sometimes thought of as an "arms race" or as "red queen dynamics." In such dynamics, the process of continued evolution in both predator and prey keeps the pair of species "standing still" like the Red Queen in Lewis Carroll's *Through the Looking-Glass*, which means that evolution allows them to continue to persist in the community (Van Valen, 1973). For space clearance rate and handling time, however, it is more of a tug-of-war, as the prey wants to push its predator's functional response down and the predator wants to pull the functional response up, creating simultaneous oppositional pressure on traits that influence the same process.

Although space clearance rate and handling time are decidedly not traits per se, they are linked to traits, so we can ask how space clearance rate and handling time, as sort of pseudo-traits, might evolve, and then guess what that would mean for the actual phenotypic traits that drive these parameters. In general, selection on space clearance rate should be positive for predators, as this would increase energy intake, resulting in increased births, remembering that births are given by the product of the functional response and the conversion efficiency (Chapter 4). The opposite would be true for prey. Selection on handling time might be expected to be negative for predators, as a lower handling time reduces the time cost of predation, allowing additional prey to be captured. If a shorter handling time comes at the expense of extracting less food from the prey, however, this might not translate into a meaningful benefit unless prey are abundant and the predator could expect another meal quickly. In contrast, the effect of handling time on the fitness of prey is less clear, as the time cost of handling itself is not paid unless the prey is actually captured by the predator, at which point the genes that influenced handling time are no longer able to be passed on. However, there might be an indirect positive effect of a higher handling time on prey via a reduction in predator abundance, thus lowering the risk of predation to the remaining prey (Livdahl, 1979). Whether this could generate selection, however, depends on whether the prey are able to reproduce before this indirect effect transpires, or whether there is some type of kin selection going on through handling time. Just the same, the indirect effect applies to all the prey, making this direction of selection potentially not very strong.

6.2 A dynamic tug-of-war

We can see how the functional response influences evolutionary fitness (the ability to add copies of oneself to the population) by examining one of our predator–prey models. The mean fitness of a population is its per capita rate of population growth because this rate is the average rate at which individuals add copies of themselves to the population. Using the MacArthur–Rosenzweig type predator–prey model that we fit to the lynx–hare data (equations (4.5) and (4.6)), we determine per capita population growth rate of the lynx by dividing both sides of the lynx equation through by lynx abundance C:

$$\frac{1}{C}\frac{dC}{dt} = e\frac{aRC^m}{1 + ahRC^m} - d \tag{6.1}$$

From this equation you can see that any trait that increases conversion efficiency e or space clearance rate a or decreases death rate d, handling time h, or interference m would increase the per capita population growth rate of the predator and therefore would be favored by selection. Likewise, we determine the hare per capita growth rate by dividing both sides of their equation through by hare abundance R:

$$\frac{1}{R}\frac{dR}{dt} = r\left(1 - \frac{R}{K}\right) - \frac{aC^{m+1}}{1 + ahRC^m} \tag{6.2}$$

Equation (6.2) shows that selection would favor the opposite effects on the functional response for prey, including a smaller space clearance rate and a higher handling time. These opposing effects on the mean fitness of predator and prey is why there is an evolutionary tug-of-war over the functional response during which prey evolve defenses and predators evolve offenses at the same time (Feldman et al., 2009; Hammill et al., 2010; Kopp and Tollrian, 2003; Ruxton et al., 2004). Note also that fitness is density dependent for both lynx and hare, indicating that fitness changes through time and therefore that understanding the evolution of traits linked to the functional response requires a dynamic approach (Abrams, 2000).

Using equations (6.1) and (6.2) parameterized from the fit of this model to the lynx–hare data from Slough and Mowat (1996) (see Chapter 4), we can determine how strong selection should be on space clearance rate and

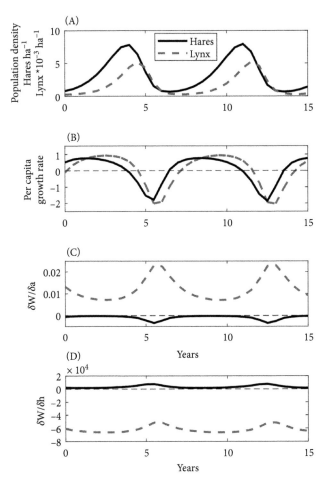

Figure 6.2 Dynamics of abundance (**A**), per capita population growth rate (**B**), and fitness gradients for space clearance rate (**C**) and handling time (**D**) for the lynx–hare system (extended from Figure 4.3). The solid black line is for hares and the dashed gray line is for lynx. Note that here I have swapped out the per capita growth rate for fitness, *W*.

handling time through time. First, re-solving the model shows that lynx and hare are expected to continue to cycle in abundance for many years (Figure 6.2A). The ups and downs in abundance through time indicate that the per capita growth rates of both lynx and hare are oscillating from positive to negative and back again (Figure 6.2B). The points where the per capita population growth rates cross the dashed line at zero growth correspond to the peaks and valleys in abundance, when growth rates shift sign and begin to increase or start to decline.

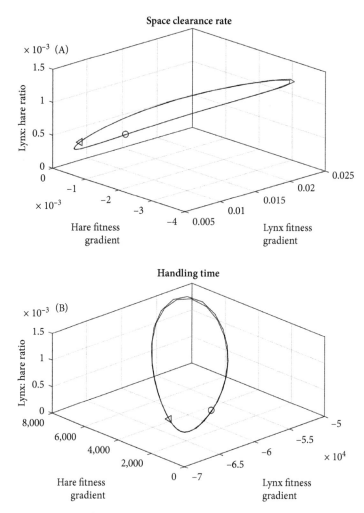

Figure 6.3 State space of fitness gradients and the ratio of lynx and hare abundance (on vertical axis) for space clearance rate (**A**) and handling time (**B**) for the lynx–hare system (after Slough and Mowat, 1996). The circle represents the start of the time series shown in Figure 6.2A; the triangle indicates the direction of movement.

6.3 Temporal variation in the strength of selection

Given that the per capita growth rate represents the mean fitness of the population, we will go one step further to understand selection on space clearance rate and handling time. Focusing first on space clearance rate, the change in fitness with a change in space clearance rate is given by the partial derivative of the per capita growth rate with respect to space clearance rate a.

Starting with equation (6.1) and using the quotient rule with $f(a) = eaRC^m$ and $g(a) = 1 + ahRC^m$, we get:

$$\frac{\partial \frac{1}{C}\frac{dC}{dt}}{\partial a} = \frac{eRC^m}{(1 + ahRC^m)^2} \tag{6.3}$$

Equation (6.3) is always positive, so a higher space clearance rate always increases the fitness of the predator (barring trade-offs, ecological pleiotropy, and long-term effects on prey abundance, see Section 6.5). That means that natural selection would have a tendency to favor predators with a high space clearance rate, as they have a higher rate of producing offspring. However, the magnitude of the positive effect of a larger space clearance rate is influenced by how many prey there are, how many predators there are, the handling time, interference, conversion efficiency, and space clearance rate itself. Thus, during the course of the lynx–hare cycles, there should be oscillations in how strongly selection is acting on predator space clearance rate. Note that here I am only considering the strength and direction of selection on functional response parameters in order to think about how we might expect the parameters to change. These equations, however, can be used in a quantitative genetics approach to model the joint changes in traits and abundance through time (eco-evolutionary dynamics). There are many other resources that demonstrate how to work on eco-evolutionary dynamics in predator–prey systems (Abrams et al., 1993; Cortez and Weitz, 2014; Lande, 1982; McPeek, 2017; Yoshida et al., 2003), so I will not cover that here.

The fitness gradient of the hare for space clearance rate is similar, since the functional response term shows up in the equation for both lynx and hare dynamics. It is also negative, since the functional response quantifies the reduction in hare population growth rate owing to foraging. Using again the quotient rule, we get:

$$\frac{\partial \frac{1}{R}\frac{dR}{dt}}{\partial a} = -\frac{C^{m+1}}{(1 + ahRC^m)^2} \tag{6.4}$$

Once again, how strongly space clearance rate affects the fitness of the hare depends on how many prey there are, how many predators there are, the handling time, interference, and space clearance rate itself.

Calculating out these "fitness gradients" (which is what equations (6.3) and (6.4) are called) through time, it appears that selection on space clearance rate (and by extension one or more of the underlying lynx and hare traits) is not constant but gets stronger and weaker through time (Figure 6.2C,D). Evolution of space clearance rate-related traits should be strongest right at the point at which both lynx and hare populations are declining at the greatest rate (Figure 6.2C). The pressure to evolve a change in space clearance rate is almost non-existent for the hare outside of this part of the cycle, and it is greatly relaxed for lynx at other times. We can see this because of the alignment of the highest (lynx) and lowest (hare) point of the fitness gradient in panel C with the lowest point of the growth rate cycles in panel B. The lynx are declining at their fastest rate just a little later than the hare. This result indicates that evolution of hare traits that lower space clearance rate should be most pronounced when predation by lynx is at its greatest level. For lynx, however, this result suggests that evolution of traits that increase space clearance rate is greatest when their foraging rate is not keeping up with their mortality rate.

We get similar but somewhat flipped outcomes for handling time fitness gradients. Following the same steps as above, the fitness gradient for lynx handling time is:

$$\frac{\partial \frac{1}{C}\frac{dC}{dt}}{\partial h} = -\frac{ea^2 R^2 C^{2m}}{(1 + ahRC^m)^2} \tag{6.5}$$

Here, an increase in the handling time reduces lynx fitness (the sign is negative) because it cuts into searching time and reduces birth rates by reducing energy uptake. In contrast, for hare, the fitness gradient is positive:

$$\frac{\partial \frac{1}{R}\frac{dR}{dt}}{\partial h} = \frac{a^2 RC^{2m+1}}{(1 + ahRC^m)^2} \tag{6.6}$$

For both lynx and hare, the strength of natural selection on handling time also should vary through time (Figure 6.2D). It turns out that selection on handling times for hare (and by extension any traits that influence handing time) also is greatest when hare populations are declining at their fastest rates. This makes sense because both space clearance rate and handling time determine foraging rate together, and the same pressures to minimize mortality when mortality is very high are operating here. However, selection to minimize handling time for lynx is greatest when lynx are increasing at their fastest rate and hare are increasing in abundance. This point in the cycle is when lynx are more constrained by their maximum rate of food intake, which is set by handling time, so reducing handling time at this part of the cycle has the greatest benefit. Together, these results suggest we should see the most pronounced evolution related to predator–prey cycles when predator and prey are experiencing the most mortality. However, we also should see changes in the predator during the growth part of the cycle for traits related to handling time (e.g., body size, digestive morphology, and feeding behaviors such as eating only the most nutritious parts of prey).

Throughout the cycles, pressure to change space clearance rate and handling time is stronger for lynx than hare, resulting in generally more and more effective predators (Figure 6.2C,D). However, equations (6.5) and (6.6) also suggest that change in space clearance rate and handling time would proceed with different dynamics during the course of a cycle. Because the fitness gradients for space clearance rate peak at the same time but have opposite signs, selection on space clearance rate would be opposite but nearly synchronous for lynx and hare (Figure 6.3A). In Figure 6.3A, we see both fitness gradients are small when the lynx:hare ratio is small, and both gradients are large when the lynx:hare ratio is large. The gradients and abundances then cycle counterclockwise (direction of the arrow) through this space with time. This outcome is a clear tug-of-war scenario, because just at the moment when selection is most strongly favoring a higher space clearance rate for predators, selection is also most strongly favoring a lower space clearance rate in prey. Thus, traits might be changing, but the functional impact on space clearance rate might be minimal. In contrast, because the fitness gradients for handling time peak at different times for lynx and hare, selection on handling time would be strongest at the opposite points of the cycle (Figure 6.3B). In Figure 6.3B, the lynx fitness gradient is large when the hare fitness gradient is small, and this occurs when the lynx:hare ratio is small as well. The gradients and abundances here cycle clockwise (direction of the arrow) through this space through time. Thus, handling time itself might vary substantially through time, generating an eco-evolutionary dynamic that changes the shape of the cycles (Yoshida et al., 2003). Together, these results show that pressure to change traits that drive space clearance rate and handling time vary through time, with abundances, and the values of other parameters, leading to a potentially broad array of evolutionary outcomes and impacts on predator–prey population dynamics. Indeed, a review of the evolution of predator–prey dynamics suggested myriad possible outcomes depending on the models, assumptions, evolving species, and whether there were links among parameters (Abrams, 2000).

6.4 Traits linked to functional response parameters

Functional response parameters reflect traits possessed by both the predator and the prey. Thus, space clearance rate and handling time do not 'belong' to predator or prey. These parameters represent ecological processes rather than specific heritable morphologies or behaviors, and thus the parameters are not really under selection themselves, but the traits that generate them may be under selection because of the demographic consequences of the functional response and any related impacts of those traits on other ecological processes. Based on our breakdown of functional response parameters and the sources of variation that we can already detect in parameters across species, we can make many predictions about which particular traits would evolve given their effects on space clearance rate and handling time. To my knowledge, however, there have been no studies showing the evolution of traits governing predator–prey interactions along with their impact on functional response parameters.

Our equation for space clearance rate (equation (3.1)) suggests that traits that influence searching velocity, the ability to detect prey or avoid detection, and any trait that confers an advantage or disadvantage in the probability of a successful attack are likely to evolve if the underlying genetic variation exists. The traits would evolve to have opposing effects between predator and prey (a tug-of-war). Offensive weaponry (e.g., claws, venom) would be promoted in predators if they increased the probability of success (p_s), and defenses (e.g., armor, chemical resistance) would be promoted in prey if they decreased that probability (Livdahl, 1979). Likewise, traits that help predators detect prey (e.g., visual acuity, chemical receptors, eavesdropping) would be promoted in predators, and traits that prevent detection (e.g., camouflage) would be promoted in prey. For example, selection has favored the evolution of fantastically large posterior median eyes in the net-casting spider *Deinopis spinosa*, presumably facilitating increased detection and probability of capture success (increasd p_s) (Stafstrom and Hebets, 2016). In contrast, the hindwing eyespots found on *Bicyclus anynana* butterflies draw the attention of potential predators such as praying mantids, such that attacking eyespots made the butterflies harder to capture (reduced p_s) (Prudic et al., 2015). Likewise, flies mimic agonistic dancing displays of the salticid spider *Phidippus apacheanus*, causing the spider to back away and be less likely to attack the fly (reduced p_a; Greene et al., 1987).

The body masses of both predator and prey also are under selection through effects on the functional response. A larger predator would generally have both a larger space clearance rate and a lower handling time, making larger size particularly advantageous for predators. The reverse would be true for prey. This latter outcome, however, would suggest that evolution would eventually lead to all prey being tiny and all predators being massive. Since that is not really what we see in nature, there must be something else influencing body size. Indeed, the unidirectional effects described here are probably never the only pressure influencing predators and their prey. Selection on traits linked to space clearance rate and handling time might also not be independent of each other.

6.5 Links among predator–prey model parameters

There are at least three ways in which selection on functional response parameters may be more complex than the simple lynx–hare example portrayed in Section 6.3. These ways include trade-offs, ecological pleiotropy, and phenotypic plasticity.

Trade-offs arise for organisms when they are faced with maintaining multiple competing priorities (Stearns, 1992). For example, a trade-off might involve allocation of a limited resource such as energy or time to different needs. A classic trade-off is that the allocation of nutritional resources to reproduction might come at a cost in growth or survival (Lee et al., 2008; Wilder et al., 2013). Thus, an organism might have higher fitness for having more offspring but then have lower fitness for not surviving long enough to have those additional offspring. For predators, an energy-based trade-off might come in the form of allocating energy to searching for prey through increased movement (V_c) or allocating it to reproduction, which in the lynx–hare model could be represented in conversion efficiency. Thus, even though a higher space clearance rate would be supported by more searching, if that increase came at the cost of energy allocated to offspring, an increase in space clearance rate might not be that beneficial. Depending on the shape of the trade-off, it could even be better for the predator to slow down and suffer a lower space clearance rate. The reverse could arise for prey, who also may need to move around to find their own food. If prey depend on their own movement to fuel reproduction, prey might actually evolve to speed up and encounter predators more often because the reproduction that would afford outweighs the increased risk of mortality. Another example of this kind of trade-off is the idea that building offensive weaponry such as larger jaws would come at the cost of allocating less energy to reproduction, lowering the conversion efficiency (Abrams, 2000). All of these features might also vary through time depending on predator and prey abundance and the value of other traits (Figure 6.2).

Ecological pleiotropy is when a particular trait has multiple, interacting effects on ecological processes (Abrams, 2000; DeLong and Gibert, 2016; Strauss and Irwin, 2004). A trait that clearly has ecologically pleiotropic effects on predator–prey interactions is body size (DeLong, 2017). As we have already seen, a larger predator can have both a larger space clearance rate and a smaller handling time, and so a larger body size has positive ecological pleiotropic effects by changing both parameters at once in the direction that increases foraging rates. For some taxa, such as mammalian carnivores (DeLong and Vasseur, 2012a), however, larger predator size can increase space clearance rate and handling time, causing antagonistic ecological pleiotropy, potentially reducing the overall benefit to the predator of becoming larger. Moreover, both predator and prey body sizes influence reproduction and mortality rates, such that changing body size will influence several aspects of their fitness at once (again in the sense of the average per capita population growth rate). Whether increasing body size has positive or negative effects on fitness, then, depends on the direction and magnitude of the net effect on all the different functions that influence fitness. It therefore can be difficult to predict whether predators or their prey should get smaller or larger given their effects on functional response parameters alone (DeLong and Belmaker, 2019). Another type of ecologically pleiotropic trait is mating displays. By making themselves detectable visually or through sound, displaying to attract mates also can attract predators that are "eavesdropping" on those displays, increasing detection by predators (d) and reproductive success of the prey at the same time. In the case of *Smilisca sila* tree frogs susceptible to predation from bats, however, calling males can use the timing of their calls to deceive predators without losing detectability by females (Legett et al., 2020), suggesting that selection sometimes can reduce the negative effects of ecological pleiotropy.

Phenotypic plasticity also may facilitate changes in space clearance rate and handling time. Plasticity can temporarily alter the parameters of the functional response as environmental conditions change, potentially limiting long-term evolutionary changes. These types of changes could be reversible behaviors for individuals (McGhee et al., 2013). For example, the protist *Lembadion* sp. can increase its overall cell volume and the size of its mouth when faced with larger prey, facilitating increased probability of success (p_s) (Kopp and Tollrian, 2003). This increase in size can be reversed by mitotic divisions back down to smaller daughter cells. In contrast, the protist *Paramecium aurelia* can become thicker and slower in the presence of the predatory flatworm

Stenostomum virginianum, reducing encounters and limiting capture success (p_s), even potentially at the cost of increased detectability (d) (Hammill et al., 2010). Finally, plasticity also can be initiated from maternal effects across generations. For example, when exposed to heightened predation risk from sparrowhawks (*Accipiter nisus*), great tits (*Parus major*) grew longer wings and had lower weight at fledging, suggesting improved predator avoidance flight abilities in the young birds (Coslovsky and Richner, 2011). Similarly, the cladoceran *Daphnia pulex* will produce neckteeth defenses if mothers are exposed to chemical signals of phantom midges (*Chaoburus* sp.) (Tollrian, 1995). This morphology could reduce predation risk by reducing the probability of attack (p_a) for this generation, but the neckteeth might not arise in the subsequent generation if predation risk is reduced. Thus, instead of evolving traits that drive the functional response up or down, predators and prey might evolve plasticity that can alter the functional response in different scenarios.

7

Optimal Foraging

This chapter is a refresher on the prey model of classic optimal foraging theory (OFT) through the lens of this book. I build on the multi-species functional response, the selection ideas, and the parameter breakdown presented in the preceding chapters to argue for how optimal foraging might arise.

7.1 Picking prey types to increase fitness

Although predation is a catastrophic negative for prey, it is only an incremental positive for predators. Each prey provides the resources predators need to maintain themselves, grow, and make new predators (Cuthbert et al., 2020b; Hartley et al., 2019). That translation of prey into predators is given by the conversion efficiency, e, described in Chapter 4. As we saw in Chapter 6, natural selection on predators will have a tendency to increase space clearance rates because that would increase energy uptake via increased foraging, thereby fueling more reproduction by the predator. By the same logic, selection on the predators would also tend to increase e, which reflects the energy contained in the prey and the cost of making new predators. Selection on predators would not change the energy contained in prey, but it could influence the way predators select prey, such that predators could increase fitness by foraging on more energetically rewarding prey. The idea that predators might forage to maximize (or just increase) their energy intake is known as "optimal foraging."

OFT emerged in the 1960s as a way to understand the complexities of animal foraging behavior and food web structure at the same time (MacArthur and Pianka, 1966). Predators seem to have many choices about what to hunt, so why do they hunt what they hunt? Some predators specialize on a limited range of prey, whereas others appear to hunt virtually any prey type that fits in their mouth. Why? OFT provides a simple answer—predators should pick prey that maximizes their net energy intake, E_n/T, where E_n is the net energy gained and time is T. Since this energy is used to produce offspring and stay alive, foraging in a way that yields more usable energy clearly has the potential

Predator Ecology: Evolutionary Ecology of the Functional Response. John P. DeLong, Oxford University Press.

to increase fitness. The only place energy shows up in equation (4.2), however, is in the conversion efficiency, where prey with higher energy content can allow predators to have more offspring.

7.2 Deriving the standard prey model optimal foraging rule

To forage optimally, a predator must be able to formulate some sort of answer to the question of whether consuming a particular prey that has been encountered will provide an improvement over their current rate of energy intake gained from whatever they are currently eating. OFT assumes that predators can gather information on the energy reward of a prey item and the time cost in handling that item (in the standard sense of a time cost to further searching). It also assumes that a predator can calculate the net rate of energy gain of the potential prey item and compare that information with its current rate of energy gain. Otherwise, how could it know if a prey item was worth it? OFT provides a mathematical way of making this decision, which presumably most predators do not actually use, but it can help us humans predict whether a predator will choose to attack a particular prey item. To do this, we return to the multi-species functional response (MSFR) from Chapter 5[1]:

$$\frac{F_{pc,1}}{T_{tot}} = \frac{a_1 R_1}{1 + a_1 R_1 h_1 + a_2 R_2 h_2} \tag{5.2}$$

$$\frac{F_{pc,2}}{T_{tot}} = \frac{a_2 R_2}{1 + a_1 R_1 h_1 + a_2 R_2 h_2} \tag{5.3}$$

Now we convert both of these equations into the rate of energy uptake rather than the rate of prey consumed. Multiplying both sides of equation (5.2) by the energy content of prey type 1, J_1 (J for joules) we get an expression for the net rate of energy gain from prey type 1 when foraging on prey types 1 and 2:

$$\frac{E_1}{T_{tot}} = \frac{J_1 F_{pc,1}}{T_{tot}} = \frac{J_1 a_1 R_1}{1 + a_1 R_1 h_1 + a_2 R_2 h_2} \tag{7.1}$$

---

[1]   In this book I use our standard functional response notation to develop the OFT prey model. This notation differs from most optimal foraging papers, which focus on encounter rates and the probability of attack. Refer back to Figure 3.1 for the relationships between encounter rates, space clearance rates, and foraging rates.

Going through the same steps we get a parallel expression for the net energy gained from prey type 2 while foraging on prey types 1 and 2:

$$\frac{E_2}{T_{tot}} = \frac{J_2 F_{pc,2}}{T_{tot}} = \frac{J_2 a_2 R_2}{1 + a_1 R_1 h_1 + a_2 R_2 h_2} \tag{7.2}$$

Again, because equations (7.1) and (7.2) have the same denominator, they can be added together to get the total energy intake from foraging on both prey types:

$$\frac{E_n}{T_{tot}} = \frac{J_1 a_1 R_1 + J_2 a_2 R_2}{1 + a_1 R_1 h_1 + a_2 R_2 h_2} \tag{7.3}$$

Assume for a moment that the predator is only consuming prey type 1. If a predator wants to maximize its net energy intake, then it needs to know whether the net energy intake is higher if it adds prey type 2 to its diet or just continues to focus on eating only prey type 1. This is determined by simply asking whether:

$$\frac{J_1 a_1 R_1}{1 + a_1 R_1 h_1} < \frac{J_1 a_1 R_1 + J_2 a_2 R_2}{1 + a_1 R_1 h_1 + a_2 R_2 h_2} \tag{7.4}$$

This expression can be simplified by moving the denominator on the right-hand side to the numerator on the left-hand side and the numerator on the left-hand side to the denominator on the right-hand side:

$$\frac{1 + a_1 R_1 h_1 + a_2 R_2 h_2}{1 + a_1 R_1 h_1} < \frac{J_1 a_1 R_1 + J_2 a_2 R_2}{J_1 a_1 R_1} \tag{7.5}$$

which simplifies to:

$$\frac{a_2 R_2 h_2}{1 + a_1 R_1 h_1} < \frac{J_2 a_2 R_2}{J_1 a_1 R_1} \tag{7.6}$$

Rearranging and simplifying again gets us:

$$\frac{J_1 a_1 R_1}{1 + a_1 R_1 h_1} < \frac{J_2}{h_2} \tag{7.7}$$

Equation (7.7) indicates that a predator should go after prey type 2 if the energy gained per time spent handling prey type 2 is greater than the current energy intake as given by the functional response and the energy contained in prey type 1, which again is referred to as the current net energy gain $\left(E_n/T\right)$. Equation (7.7) is the classic OFT prey selection decision criterion (Charnov, 1976).

There are two important things to notice here. First, remember that the maximum foraging rate, or asymptote of a saturating functional response, is $1/h$, so J/h is the asymptotic rate of energy intake. Second, the density of prey type 2 is not in equation (7.7), so according to this model it does not factor into the decision to eat it (Charnov, 1976). It only matters whether the net energy intake rate foraging on prey type 1 falls below the energy asymptote of the second prey type (Figure 7.1). The density at which prey type 1 becomes

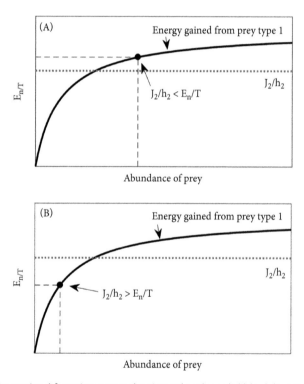

Figure 7.1 The optimal foraging prey selection rule. The solid black line is the energy gain (energy * functional response) for prey type 1. In (**A**) the net energy intake foraging on prey type 1 (E_n/T, solid black dot) is greater than the expected energy intake for an individual of prey type 2 (J_2/h_2, dashed gray line). Therefore, the predator should not eat type 2. In (**B**) the reverse occurs, and so the predator can increase its overall energy intake by adding prey type 2 to its diet. The decision to add another prey type to the diet varies depending on the abundance of prey (on the x axis) that the predator has already included in its diet, but not the abundance of prey type 2.

rare enough that it is worth eating prey type II can be found by solving for R_1 in equation (7.7):

$$R_1 = \frac{J_2}{a_1 (J_1 h_2 - J_2 h_1)} \tag{7.8}$$

OFT suggests that predators can actively choose to attack or not attack particular prey based on knowledge of the benefit of that prey. This means that the probability of attack (p_a) component of the space clearance rate (equation (3.1)) must switch from 0 (do not ever attack) to 1 (always attack) as the net energy intake declines from above the expected energetic value of the prey being considered (J/h) to below that expected value (Figure 7.2). In classic OFT, this switch from 0 to 1 is a step function, meaning an abrupt switch perfectly tuned to maximize fitness (Figure 7.2). In reality, predators have imperfect information about the energetic value of prey, their current net energy intake, as well as uncertainty about the future availability of any prey type including the ones already in the diet. Thus, an alternative expectation,

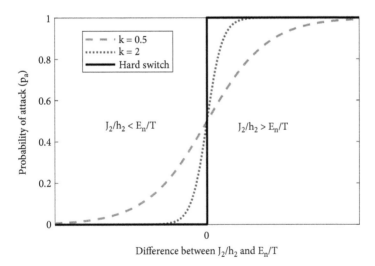

Figure 7.2 The switch in the probability of attack (p_a) as the difference between the current net energy intake given prey type 1 (E_n/T) moves from being greater than the expected energy gain of prey type 2 (J_2/h_2) to being less than prey type 2 (moving left to right). In classic OFT, the switch is predicted to be a step function; that is, the predators perfectly align their prey preferences with the actual energetic rewards. In reality, imperfect information, individual variation, and the uncertainty around finding the next prey item could all pay a role softening the transition to varying degrees (dashed orange and dotted blue lines).

one that matches prey-switching behavior more closely (Krebs et al., 1977), is that predators display a gradual shift, slowly adding more of prey type 2 as its benefit increases. This can be described with a logistic function:

$$p_a = \frac{1}{1 + e^{-k\left(\frac{E_n}{T} - \frac{J_2}{h_2}\right)}} \tag{7.9}$$

In equation (7.9), the probability of attack is 0.5 when $\frac{E_n}{T} = \frac{J_2}{h_2}$ and transitions to 0 when the current energy intake rate $\left(E_n/T\right)$ is relatively high and to 1 when E_n/T is relatively low. The slope of this transition is given by the parameter k. Although just a phenomenological representation, this function is what OFT predicts should happen (Figure 7.2).

7.3 OFT remains useful and needs further testing

OFT theory (the prey model, as this particular type of OFT is known) has been tested many times (Sih and Christensen, 2001). Some tests have been qualitative (*yes* or *no* eat a particular type of prey), and some have been quantitative (choose to include this type of secondary prey at this density of the preferred prey). In the aggregate, the number of studies supporting OFT seem to outweigh the number that refute it (Sih and Christensen, 2001). Furthermore, OFT seems to work better for non-mobile than mobile prey types (Sih and Christensen, 2001), suggesting the need to gather more information and evaluate behavioral details of prey when testing OFT. Somewhat surprisingly, even though OFT has been developed through the lens of the functional response, it appears rare for studies testing OFT predictions to actually measure functional responses to help generate those predictions. If, for example, a researcher used a behavioral measure of handling time instead of that estimated from a functional response, they might overstate the energetic value of a prey type and thus make an incorrect prediction about when a predator should include that prey type in its diet. This could be one reason why the empirical support for OFT is inconsistent.

Although OFT is no longer in its heyday of interest and has been criticized heavily (Pierce and Ollason, 1987), it remains a foundational concept in ecology. It is still not clear why the theory works well in some cases and not well in others. Yet, the logic of OFT—that predator foraging behavior would be under selection to increase energy intake—is still sound, and the theory may yet suggest new ways in which to interpret the foraging behavior of predators. For example, imagine a golden eagle (*Aquila chrysaetos*) cruising some montane grassland, encountering prey, and deciding whether to attack. According to OFT, each time the eagle spots something it recognizes as prey, it makes a decision to attack or not based on the prey's profitability. It might make that decision based on how large a prey item it is, since that might be related to the energy gain it could expect to get. It could also evaluate the time cost (or even the risk of injury) to attacking and consuming that prey item. If the eagle knows what types of prey it might pass up while it is consuming this prey item, it could make a decision that maximizes its energy intake. Interestingly, golden eagles, like many raptors, consume an enormous variety of prey types, from lizards to deer (Carnie, 1954; Hector, 1985; Stahlecker et al., 2009). This suggests that most prey items qualify as energetically rewarding given the bird's current condition. The inclusion of a wide variety of small prey in the diet of this very large bird, when they could be consuming prairie dogs or ducks, suggests that the net energy intake of a predator like the golden eagle is perpetually low, and that only in rare, flush situations might they actually start becoming choosy about what they eat. OFT suggests, then, that predators are generalists because foraging (and thus net energy gain) rates are low most of the time. This observation is in conflict with another hypothesis, however, which is that predators are often able to reach satiation and therefore are not time-limited for foraging (Jeschke, 2007). This possibility suggests that many predators should be specialists, not wanting to incur unnecessary handling time in acquiring more energy. Returning to Chapter 4, however, when examining the observed densities of hare and the estimated functional response of lynx on hare given the fitted dynamics, it appears that lynx foraging rates are much lower than they could be if hare were more abundant (Figure 7.3). Thus, the lynx–hare system looks more like Figure 7.1B than 7.1A, and lynx therefore could increase their net energy intake by adding other types of prey to their diet, which they do at least in some situations (Chan et al., 2017). Similarly, densities of prey for the predatory intertidal whelk, *Haustrum scobina*, seem low enough in natural settings to make it challenging to detect the saturation prey levels (Novak, 2010), suggesting that these predators are generally able to eat more than the environment provides.

Finally, OFT suggests that MSFRs are dynamic constructions, changing as the abundance of the most energetically rewarding prey changes. As the most energetically rewarding prey becomes rare, more and more low-reward prey are added to the diet, expanding the MSFR to include more species and potentially reducing realized diet specialization (Araújo et al., 2011). However, if we take an MSFR as a fixed construct, then predators would

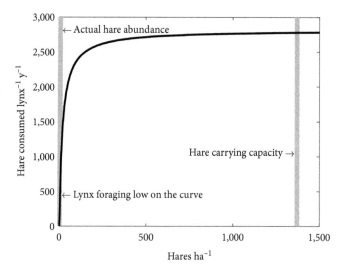

Figure 7.3 The functional response of lynx foraging on hare according to the fitted dynamics in Chapter 4. When hare are at their carrying capacity, lynx would forage on more than 2,500 hares per year. Given the observed range of actual hare densities over a cycle, however, lynx forage at a fraction of that rate, suggesting that lynx are perpetually foraging at rates that would keep E_n/T low, making many other prey types energetically rewarding given OFT.

be predicted to consume some prey even if doing so lowered the overall net energy intake of the predator. This has been referred to as "suboptimal" foraging (Vallina et al., 2014) because the model suggests that predators must consume prey types that lower E_n/T. It is not clear the degree to which predators actually lower their energy intake rate by foraging on low-reward prey, because it is so difficult to know what their net energy intake would be without it. In one case, however, walleye (*Sander vitreus*) had lower energy intake when foraging on larval carp (*Cyprinus carpio*) and the water flea *Daphnia pulex* together than when foraging only on larval carp (Czesny et al., 2001). If foragers simply must consume additional prey types according to an MSFR, however, then predators would be expected to generally have lower net energy intake rates, and thus lower fitness, in areas with greater prey diversity, because some of that diverse prey base will be low-reward. This possibility makes sense unless the most energetically rewarding prey are perpetually rare, making everything else roughly equally rewarding.

To take this discussion to its logical conclusion, an MSFR that prevents suboptimal foraging could be written out based on the OFT functions described above. Given that the processes leading up to foraging events can be described by combining equations (3.1), (5.4), and (7.9), where *a* in equation (5.4) has been replaced by equation (3.1) and the p_a in equation (3.1) has been replaced by equation (7.9), we can write:

$$\frac{F_{pc}}{T_{tot}} = \frac{\sum_i \left(p_s \left(\frac{1}{1+e^{-k\left(\frac{E_n}{T} - \frac{E_i}{h_i}\right)}} \right) d\sqrt{V_c^2 + V_i^2}\, R_i \right)}{1 + \sum_i \left(p_s \left(\frac{1}{1+e^{-k\left(\frac{E_n}{T} - \frac{E_i}{h_i}\right)}} \right) d\sqrt{V_c^2 + V_i^2}\, R_i h_i \right)} \tag{7.10}$$

This OFT-based MSFR is built on behavioral mechanisms underlying the foraging process, but might be a bit hard to parameterize.

8

Detecting Prey Preferences and Prey Switching

In this chapter I consider the question of whether predators switch their preference for different types of prey as those prey change in abundance. There are numerous experiments in the literature focusing on this, but generally they have focused on a simplified analysis that ignores the functional response. Here I show why the functional response is crucial for understanding prey choice.

8.1 Prey selection in the presence of alternate prey

Pairwise predator–prey functional responses describe the foraging rate of a predator on a particular prey in isolation. Most predators in nature, however, forage in a community that contains multiple prey types (Chapter 5). The presence of alternative prey could alter the functional response relative to the single-prey situation in two general ways beyond having to add up the time for handling all prey types as we did with the multi-species functional response (MSFR). First, interactions among prey could alter their movements (V_r), altering the rate of encounters, and by virtue of changing movement, alter the prey's detectability as well (d). Second, predators could choose among prey types based on the relative (dis-)advantage of different items on the "menu," altering their likelihood of attack (p_a). Recalling that all of these factors are components of the space clearance rate (equation (3.1)), the presence of alternative prey could alter the functional response and thus the foraging rates relative to that predicted by the single-prey scenario.

Change in prey selection as the abundance of a specific prey type changes is known as "prey switching" (Lawton et al., 1974; Murdoch, 1969). Prey switching generally refers to the idea that predators will forage more on a particular prey type as it becomes more abundant, but if the predators forage more on a prey type as it becomes more rare then it is negative or reverse prey switching. A similar idea is that predators switch based on the total availability

Predator Ecology: Evolutionary Ecology of the Functional Response. John P. DeLong, Oxford University Press.
© John P. DeLong 2021. DOI: 10.1093/oso/9780192895509.003.0008

of prey, referred to as "rank" switching (Baudrot et al., 2016). A predator might switch prey if, for example, it seeks to add prey that provide specific nutrients (Belovsky, 1978), contain more protein (Schmidt et al., 2012), or that are less dangerous (Tallian et al., 2017). It is also possible that predators get better at finding and handling specific prey types the more they eat them (i.e., the functional response increases with practice), so predators might increasingly forage on whatever they are most experienced with (Bergelson, 1985). Either of these possibilities requires that different potential prey types must be discernable by the predator and that the predator is able to make different decisions based on that information (Morozov and Petrovskii, 2013). A classic study on the backswimmer *Notonecta glauca* showed that when given variation in the relative frequency of two prey types—mayfly larvae (*Cloeon dipterum*) and the isopod *Asellus aquaticus*—the backswimmers ate prey in proportion to their availability at first (Lawton et al., 1974). After 8 days, however, the backswimmers took the more abundant prey more than expected. This outcome suggests that after some experience assessing the situation, and presumably gaining experience hunting in the arenas, the backswimmers altered some component of their space clearance rate that allowed them to achieve higher predation rates. Indeed, the backswimmers showed greater capture success on isopods when reared for a week on a higher proportion of isopods. In another classic example, guppies (*Poecilia reticulata*) switched from preying on flies (*Drosophila*) when they were more abundant to preying on tubificid worms when they were more abundant by spending more time searching in the habitat with the most abundant prey type (Murdoch et al., 1975). In this case, then, prey "switching" really means foraging patch switching, and not necessarily just a switch in functional response parameters in the same context.

8.2 Detecting prey switching

However it arises, detecting prey switching requires being able to measure differences in the rate of foraging on particular prey between scenarios where a particular prey type is in isolation and scenarios where that prey is in a community of other prey types. Most typically this is done by comparing the proportion of prey types consumed with the proportion expected to be consumed. The actual proportion of prey type *i* in the diet (d_i) is:

$$d_i = \frac{n_i}{\sum_j n_j}$$

(8.1)

where n_i is the number of prey type i consumed, and the denominator is the sum of the number eaten of all prey types j. The most basic null expectation is that predators consume prey type i in proportion p_i to its availability, meaning:

$$d_i = p_i = \frac{R_i}{\sum_j R_j} \qquad (8.2)$$

where R_i is the number of type i available, and the denominator is now the sum of the number available of all prey types j (Chesson, 1983; Cock, 1978). In any given foraging scenario, $n_i \leq R_i$, since a predator cannot eat more than is available. Therefore we can write $n_i = \alpha_i R_i$, where α_i is between zero and one. Substituting into equation (8.1), we get:

$$d_i = \frac{\alpha_i R_i}{\sum_j \alpha_j R_j} \qquad (8.3)$$

which reduces to equation (8.2) whenever all α are equal and therefore cancel out. Thus, one traditional way of determining whether a prey type is preferred or selected is whether the αs differ among prey types. In experiments, the simplest way of looking at this is by calculating them as $\alpha_i = \frac{n_i}{R_i}$.

As with most functional response experiments, prey are reduced in abundance during the experiment. The standard solution to this problem is to use a selection index known as Manly's α. This widely used index (e.g., Chesson, 1983; Cock, 1978; Jaworski et al., 2013; Klecka and Boukal, 2012) compares the natural log of number of a prey type i eaten as a fraction of type i offered relative to the sum of the natural logs of the fractions eaten of all prey types offered:

$$m_{\alpha i} = \frac{\ln\left(\frac{R_i - n_i}{R_i}\right)}{\sum_j \ln\left(\frac{R_j - n_j}{R_j}\right)} \qquad (8.4)$$

where R_i is again the number of prey type i offered, n_i is the number of prey type i eaten, and there are j prey types. Manly's α evaluates to 0 when none of prey type i are eaten (because $\ln(1) = 0$) and 1 when none of the other prey types are eaten.[1] When prey are eaten in equal proportions, Manly's α is 0.5. Substituting $n_i = \alpha_i R_i$ into equation (8.4), Manly's α becomes:

---

[1] One problem with this index is that whenever all of any prey type are eaten, the index is not defined, because $\ln(0)$ is $-$infinity. This means that one must be careful to not run experiments to the point where all prey are consumed.

$$m_{\alpha i} = \frac{\ln\left(1 - \alpha_i\right)}{\sum_j \ln\left(1 - \alpha_j\right)} \tag{8.5}$$

which clarifies that the two αs are not the same, Manly's α nevertheless is a function of the original αs, and that whenever the original αs are all equal, it resolves to 0.5 in a two-prey system (or 0.33 in a three-prey system, etc.). Thus, a difference between the observed Manly's α for prey type i and 0.5 (in a two-prey system) indicates a "preference" for or against that prey type. If the predator's α for a particular prey type increases as that prey type increases in relative abundance, then the predator is said to display positive prey switching. If the α increases as particular prey type becomes a lower fraction of the prey, then the predator is showing reverse prey switching. Both of these switching behaviors indicate that the predator is consuming more or less than expected, suggesting a preference. Chesson (1983) made clear, however, that preference really means any number of different mechanisms (i.e., encounters, detections, decision to attack) that could change foraging rates, not just the innate preference of a predator for a particular prey.

8.3 Null expectations from the functional response

If we are looking to understand the shift in prey selection owing to the presence of alternate prey, however, the relative abundance of a prey type is not a good indicator of its expected frequency in a mixed-prey diet (Kalinkat et al., 2011). That is because we generally expect the predator to eat some fraction of what is available that is determined by the MSFR, not necessarily everything that was offered. This expectation arises because any prey that have different functional responses with the predator of interest will automatically show α values that differ from each other even when predators are foraging exactly as they would in single-prey scenarios. Thus, despite being widely used, comparisons of the proportion of prey in the diet with the proportion available cannot reliably tell us about preference. For example, if a predator was fed equal numbers of prey types i and k but the space clearance rate on i was twice that on k, we would expect that $\alpha_i > \alpha_k$ and that Manly's $\alpha > 0.5$, both of which would indicate a preference for i even though the predator is eating the amount expected for each type of prey with no preference for either. Thus, neither of the classic null expectations (all αs are equal or that Manly's $\alpha = 0.5$) adequately represent the case of neutrality with respect to prey selection. This fact was pointed out long ago (Cock, 1978; Lawton et al., 1974), although the importance of using the functional response as a baseline seems to have faded over the years (Chesson, 1983; Hellström et al., 2014; Joyce et al., 2019; Kalinkat et al., 2011; Klecka and Boukal, 2012; Krylov, 1988). This might be because of the requirement of running both functional response and prey choice experiments and the challenge of working with MSFRs even in laboratory settings (Chapter 5). Nonetheless, it is critical going forward to understand how prey preference depends on the MSFR baseline.

To derive the MSFR expectation for neutral prey selection, we simply plug in the MSFR prediction for the number of prey eaten (F_{pc}) for the number of prey available (R) in equation (8.2). Recall from Chapter 5 and equations (5.2) and (5.3), the total number of prey type 1 consumed in time T_{tot}, given also the consumption of prey type 2, is:

$$F_{pc,1} = \frac{a_1 R_1 T_{tot}}{1 + a_1 R_1 h_1 + a_2 R_2 h_2} \tag{8.6}$$

Accordingly, the total number of prey type 2 consumed in time T_{tot} given the consumption of prey type 1, is:

$$F_{pc,2} = \frac{a_2 R_2 T_{tot}}{1 + a_1 R_1 h_1 + a_2 R_2 h_2} \tag{8.7}$$

Following equation (8.1), the expected proportion of prey type 1 in the diet would therefore be:

$$p_1 = \frac{F_{pc,1}}{F_{pc,1} + F_{pc,2}} \tag{8.8}$$

or:

$$p_1 = \frac{\frac{a_1 R_1 T_{tot}}{1 + a_1 R_1 h_1 + a_2 R_2 h_2}}{\frac{a_1 R_1 T_{tot}}{1 + a_1 R_1 h_1 + a_2 R_2 h_2} + \frac{a_2 R_2 T_{tot}}{1 + a_1 R_1 h_1 + a_2 R_2 h_2}} \tag{8.9}$$

Noticing that the T_{tot} and the denominator of each functional response cancel out, this expression reduces to:

$$p_1 = \frac{a_1 R_1}{a_1 R_1 + a_2 R_2} \tag{8.10}$$

This somewhat surprisingly returns us to something that looks like equation (8.3), but here the proportion expected is not strictly given by their abundances but also by their single-species space clearance rate. If the space clearance rates of the two prey types are the same, then the proportional expectation works, but it does not work if the space clearance rates are different. Equation (8.10) also suggests that handling time does not play a role in setting the proportion of a prey type in the diet, even though handling time can strongly influence foraging rates at higher densities. Why would handling time not matter here?

As an example, Figure 8.1 shows functional responses of deutonymphs (second larval stage) of the predaceous mite *Phytoseiulus persimilis* foraging on larvae and deutonymphs of the red spider mite (*Tetranychus urticae*) (Cock, 1978). The functional response for larvae is much higher than the functional response for deutonymphs. This difference arises primarily owing to a difference in handling time and not a difference in space clearance rate (Figure 8.1). The higher functional response suggests that whenever the predator is offered more than 10 or so prey, one might expect that the predator would eat about two times more larvae than deutonymphs. However, they

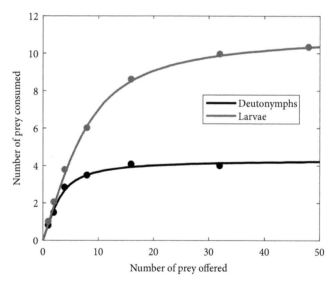

Figure 8.1 Functional responses of predaceous mites (*Phytoseiulus persimilis*) on larvae and deutonymphs of red spider mites (*Tetranychus urticae*) (data digitized from Cock, 1978). The higher functional response on larvae is caused by a lower handling time (2.15 h for larvae versus 5.53 h for deutonymphs), but the space clearance rates are very similar between the two prey types (0.13 for larvae versus 0.12 for deutonymphs arenas per predator per hour).

would not eat more larvae unless they actually preferred them. Recall from Chapter 2 that the search time is the total time minus the handling time ($T_s = T_{tot} - aRT_sh$), and that the type II functional response is just the type I multiplied by the proportion of time spent searching ($f_{pc} = aR\frac{T_s}{T_{tot}}$). The same is true for a type II MSFR, where the search time discount is the same for all prey types. Starting from equation (8.8) again:

$$p_1 = \frac{a_1 R_1 \frac{T_s}{T_{tot}}}{a_1 R_1 \frac{T_s}{T_{tot}} + a_2 R_2 \frac{T_s}{T_{tot}}} \qquad (8.11)$$

and with the $\frac{T_s}{T_{tot}}$ canceling out, this brings us right back to:

$$p_1 = \frac{a_1 R_1}{a_1 R_1 + a_2 R_2} \qquad (8.10)$$

Since the total handling time is the cause of the loss of search time for each species, it cancels out and does not influence the expectation.

To evaluate whether prey depletion changes things, we could use the Roger's Random Predator (RRP) equation (equation (2.15)) to determine the number of prey expected to be consumed in the two-prey scenario in equation (8.7) (Cock, 1978; Lawton et al., 1974). The two-prey version of this is again:

$$F_{pc,1} = R_{o,1}\left(1 - e^{a_1\left(h_2 F_{pc,2} + h_1 F_{pc,1} - T_{tot}\right)}\right) \qquad (5.6a)$$

$$F_{pc,2} = R_{o,2}\left(1 - e^{a_2\left(h_2 F_{pc,2} + h_1 F_{pc,1} - T_{tot}\right)}\right) \qquad (5.6b)$$

Notice that, as before, the number of prey consumed cannot be isolated on the left-hand side and the equations contain the number of prey consumed of the alternate prey type on the right-hand side. This means that one must use numerical solutions to estimate the number of each prey type eaten given the single-species functional response parameters. The Lambert Random Predator equation can isolate the prey eaten on the left-hand side for the single-species case (Bolker, 2011). Following the same approach (see Box 5.1), the two-species Lambert$W$ version can be identified, but, as with the RRP, the alternative prey type is still present, meaning that we still have to use numerical solvers to estimate the predicted number of prey eaten. Given the complexity of the two-species Lambert Random Predator formula, we might as well stick with the RRP version. Plugging equations (5.6) into equation (8.8), then, yields:

$$p_1 = \frac{R_{o,1}\left(1 - e^{a_1\left(h_2 F_{pc,2} + h_1 F_{pc,1} - T_{tot}\right)}\right)}{R_{o,1}\left(1 - e^{a_1\left(h_2 F_{pc,2} + h_1 F_{pc,1} - T_{tot}\right)}\right) + R_{o,2}\left(1 - e^{a_2\left(h_2 F_{pc,2} + h_1 F_{pc,1} - T_{tot}\right)}\right)}$$

(8.12)

which I think is nicer rearranged to:

$$p_1 = \frac{1}{1 + \frac{R_{o,2}\left(1 - e^{a_2\left(h_2 F_{pc,2} + h_1 F_{pc,1} - T_{tot}\right)}\right)}{R_{o,1}\left(1 - e^{a_1\left(h_2 F_{pc,2} + h_1 F_{pc,1} - T_{tot}\right)}\right)}}$$

(8.13)

Remembering that $T_s = h_2 F_{pc,2} + h_1 F_{pc,1} - T_{tot}$, this reduces to:

$$p_1 = \frac{1}{1 + \frac{R_{o,2}\left(1 - e^{a_2 T_s}\right)}{R_{o,1}\left(1 - e^{a_1 T_s}\right)}}$$

(8.14)

which again shows that handling time differences are not expected to influence the proportion of prey types in the diet. Thus, when assessing prey preferences or prey switching, the handling time does not matter, but when trying to predict the foraging rate or amount of foraging in an MSFR, the handling time does matter.

To illustrate this directly, the proportion of larvae consumed by predaceous mites (the same system as shown in Figure 8.1) in mixed-prey trials is shown in Figure 8.2 (Cock, 1978). The predaceous mites appear not to show any preference for larvae or deutonymph red spider mites, given that a regression of the observed proportion on the predicted proportion includes all versions of the null expectation. I used both equations (8.10) and (8.13) as null expectations, and it is clear that they are exactly the same, which again shows that handling time is not relevant to the prediction of proportions. The null expectation, however, is that the proportions would be about the same, since the space clearance rates are nearly identical between the prey types (Figure 8.2).

It is also worth noting that if predators alter their foraging such that they consume more or fewer than expected, this implies that there is some multiplier on the space clearance rate in the multi-prey scenarios:

$$d_1 = \frac{\beta_1 a_1 R_1}{\beta_1 a_1 R_1 + \beta_2 a_2 R_2}$$

(8.15)

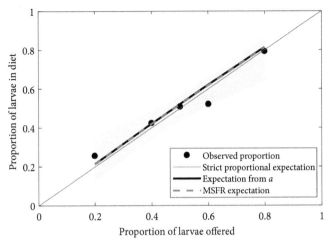

Figure 8.2 Proportion of larvae in the diet of deutonymph *Phytoseiulus persimilis* (data from Cock, 1978). Across trials where larvae made up an increasing fraction of 40 prey items offered, we would expect spiders to increase the relative amount of larvae in the diet. The traditional null expectation is that predators would consume prey in proportion to their availability in the environment. Two MSFR-based null expectations are based on space clearance rate alone (solid blue line) or estimated from the full MSFR (dashed orange line). It is clear that these two expectations are exactly the same, indicating that handling time is not necessary for predicting the proportion. The shaded gray area is the 95% confidence intervals of a linear regression of observed proportion on predicted proportion.

where the βs are the proportion of prey eaten but relative to that expected by the functional response rather than that offered, which means that unlike the original αs, they can be > 1, but $\beta_1 a_1 \leq 1$ because it is still a proportion.

8.4 Null expectations for Manly's α

Given that the MSFR is essential for predicting the proportion of a prey type in the diet, the expected number consumed from the MSFR would make a more accurate null expectation for Manly's α as well. In a two-prey system, having no preference is traditionally given by $\alpha = 0.5$. We can use the Manly's α formula to generate an MSFR-based expectation by predicting the value of Manly's α when the predator consumes exactly that amount predicted by the MSFR:

$$m_{\alpha i} = \frac{\ln\left(\frac{R_i - F_{pc,1}}{R_i}\right)}{\sum_j \ln\left(\frac{R_j - F_{pc,j}}{R_j}\right)} \tag{8.16}$$

This equation again requires the use of either version of the random predator equation because here we need the number of prey items of each type consumed.

As an example, I reanalyzed the data from a study on wolf spiders (*Pardosa pseudoannulata*) in rice fields foraging on both planthoppers (*Nilaparvata lugens*), an herbivore of rice, and on mirid bugs (*Cyrtorhinus lividipennis*), which are an egg predator of the planthopper (Heong et al., 1991). In single-prey trials, the wolf spider had a larger space clearance rate and lower handling time when foraging on mirid bugs than when foraging on planthoppers (Figure 8.3). Given the differences in the functional response, we would not expect the same number of each prey type to be consumed even when the same number of each prey type was offered. Using the null-MSFR predictions (given the single-species functional responses) for each prey type from equation (8.10), the expected Manly's α is 0.21, much lower than the 0.5 null expectation that does not account for differences in the functional response (Figure 8.4). This value does not change with the ratio, as Manly's α

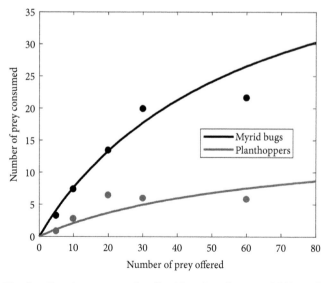

Figure 8.3 The functional response of wolf spiders foraging on mirid bugs has a higher space clearance rate (0.89 versus 0.24 arenas per predator per day) and a lower handling time (0.02 vs. 0.06 days) than planthoppers. Data digitized from Heong (1991).

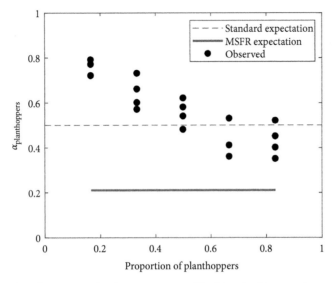

Figure 8.4 When foraging on both prey types, wolf spiders in this experiment (data digitized from Heong, 1991) clearly "preferred" planthoppers at all densities. This conclusion differs from the one drawn from a proportional abundance expectation of 0.5. In the standard approach, it appears that wolf spiders only show a preference for planthoppers when they are rare. In the MSFR approach, spiders always show a preference for planthoppers, but that preference increases as the planthoppers become rarer.

is not sensitive to the number of prey offered (hence its use for managing the problem of prey depletion).

The observed Manly's α data values (with respect to planthoppers) for their spiders were relatively high when planthoppers were rare and declined as planthoppers became relatively more frequent (Figure 8.4). Using the traditional null expectation, this outcome suggests that the spiders showed reverse prey switching, preferring planthoppers when they were rare but taking them in proportion to their abundance when common. The MSFR expectation suggests a slightly different interpretation. Spiders always preferred planthoppers but more so when they were rare. There are quite a few prey selection experiments in the literature, generally using the proportion of prey available or a Manly's α of 0.5 as a null expectation (Klecka and Boukal, 2012). This analysis suggests that many of those works could be reanalyzed with a different null expectation if suitable data on the predator's functional response are available.

In any case, consuming more planthoppers than expected in this experiment must come from an increase in encounters, detection, attack probability, or success probability; that is, an increase in the functional response on planthoppers between the single-prey and the two-prey scenarios. It is not clear what exactly happened in this experiment, but one possibility is that planthoppers simply moved around more as a result of interactions with mirid bugs, but that this effect decreased as the planthoppers became less rare. What this would mean is that the encounter rates had gone up in the presence of the other species, increasing the space clearance rates via the effect on velocity (V_r) (equation (3.1)). Alternatively, planthoppers may contain important rare nutrients. To obtain those nutrients, spiders may prioritize planthoppers to a greater extent if they encounter them less frequently, which would in effect increase the probability of attack (p_a), also increasing the space clearance rate on planthoppers. Either way, deviations from the null expectation suggest changes in the behaviors and interactions that set the functional response parameters and lead to shifts in the foraging behavior.

9
Origin of the Type III Functional Response

In this chapter I review the many ways that functional responses may show a sigmoidal shape rather than the simpler asymptotic shape. I also raise some concerns with the standard type III model and offer an alternative.

9.1 What generates a type III functional response?

As described in Chapter 2, the type III functional response is one in which the curve is relatively shallow at low densities and gets steeper as prey levels increase, generating a sigmoidal shape instead of the typical, decelerating asymptotic shape. There is no clear consensus in the literature about whether type II or type III functional responses are more common. Furthermore, although Jeschke et al. (2004) argued that type I functional responses are exclusive to filter feeders, it is still unclear why a type II or type III functional response might arise in different scenarios or among different predator–prey pairs (Dunn and Hovel, 2020; Jeschke et al., 2002). Regardless of their relative frequency, there appear to be empirically sound cases of type III functional responses. Furthermore, there are important stabilizing and ecosystem consequences of type III functional responses (Anderson et al., 2010; Daugaard et al., 2019; Hammill et al., 2010; Oaten and Murdoch, 1975; Uszko et al., 2015), indicating a need to understand what they represent in terms of the overall evolutionary ecology of functional responses.

The standard type III model characterized by an exponent on prey levels (equation (2.7)) does not, to my knowledge, have a mechanistic derivation. However, Real (1977) presented a derivation analogous to enzyme kinetics that generated an exponent, n, on prey levels, where n is the number of "binding sites." Since the exponent was presented as an integer and is analogous to the number of prey items needed to be consumed for a predator to reach its maximum efficiency, it seems incompatible with the reality that most fitted Hill exponents are not integers and only slightly different from

Predator Ecology: Evolutionary Ecology of the Functional Response. John P. DeLong, Oxford University Press.
© John P. DeLong 2021. DOI: 10.1093/oso/9780192895509.003.0009

one, indicating that the model does not quite work. However, it is clear that a type III functional response may arise if space clearance rate changes as some function of prey levels (equation (2.8)). Since we know that space clearance rate can be broken down into its component parts representing aspects of the foraging process (equation (3.1)), we can identify several hypotheses about the mechanisms that might generate a type III functional response by identifying ways in which the components might be functions of prey levels.

Recall equation (3.1), where we expanded space clearance rate with the expression $a = p_s p_a d \sqrt{V_c^2 + V_r^2}$. This expression suggests there are at least five aspects of space clearance rate (i.e., component parameters) that could increase with prey levels and therefore could generate a sigmoidal functional response. First, the probability of success (p_s) could increase with prey levels. This could happen if prey become less wary as prey levels increase, perhaps owing to distractions or impaired defense behaviors that arise with more prey–prey encounters. This also could occur if the per capita risk to prey is lower by virtue of there being many other prey the predators could choose to consume. Second, the probability of attack (p_a) could increase with prey levels. This possibility is the classic "prey switching" hypothesis, where predators begin to hunt a particular prey type more actively as they become more abundant, which could arise for a wide variety of reasons (Real, 1977). Such an outcome could happen if there is a positive feedback between successful hunts and further pursuit of specific prey. It also could happen if the cost of hunting is higher than the expected gain from hunting rare prey, meaning that if hunting is energetically expensive, predators might be more likely to hunt when there are more prey around so that they can pay the hunting costs and still have energy left over for other needs (Kiørboe et al., 2018). Third, prey also could become increasingly detectable as they become more abundant, perhaps if some individuals are forced into riskier locations by other prey or through an increase in movements that make prey more obvious to the predator. Alternatively, predators could develop an improved ability to detect prey with repeat exposures by refining a search image of the prey (Bond and Kamil, 1998; Pietrewicz and Kamil, 1979). Fourth, predators could increase their searching velocity as prey become more available. This might arise if predators use information from encounters with prey to determine when it is most rewarding to search for prey and then searching more actively when prey are more abundant. Fifth, prey may increase search velocity for their own prey as they become more abundant, increasing their encounters with predators. This could arise if prey need to travel greater distances in response to competition from other prey. Finally, it may be that the relative velocity takes on a different expression than the root sum of squares. This expression essentially says that the outcome of encounters looks like what would happen if predator and prey were gas molecules, even if they really are not behaving like gas molecules. Even though moving faster as prey abundance increased could create a type III functional response, it is possible that movement strategies would vary with prey abundance such that the root sum of squares expression is not appropriate at all prey levels. Thus, there could be another expression governing this process that would change with prey abundance.

Any of these mechanisms, alone or in combination, could generate a type III functional response. They sometimes represent a type of behavioral response to the abundance of prey, which would require, at the minimum, the ability for the predator to gather information about prey availability. Other mechanisms arise through changes in the behavior of the prey. Some of these possible mechanisms may not be likely for certain foraging strategies, such as sit-and-wait predators, who have zero searching velocity and thus would not move more when prey become abundant. However, some predators may have the ability to utilize flexible foraging strategies, such as expending more energy on searching when there are more prey about. Type III functional responses also could arise in different ways in different systems. Numerous studies have reported on cases in which the type III response is seen for one functional response but not for another, related functional response that differs in some specific way (Chan et al., 2017; Daugaard et al., 2019; Hammill et al., 2010; Hossie and Murray, 2016; Morozov, 2010; Sarnelle and Wilson, 2008). Given the different mathematical ways that components of space clearance rate could vary with prey abundance, it seems that there could be a range of sigmoidal shapes that would be consistent with what we call a type III functional response. That is, we now classify sigmoidal curves as one type of functional response (the type III), but the different ways in which a sigmoidal functional response could arise create the possibility that there are several "types" of type III, where foraging related behaviors vary in a variety of ways with prey (or even predator) abundance (Okuyama, 2012).

There is very little research yet that would allow us to discern which behavioral mechanisms might be at play in the generation of type III functional responses. However, a few studies suggest what behavioral mechanisms might be important. For example, Bergelson (1985) found that dragonfly nymphs (*Anax junius*) became more efficient as they foraged more on a particular prey type, increasing their orientation toward and pursuit of *Tubifex* worms as they became more abundant. This outcome is consistent with both an increase in p_a and p_s with the abundance of worms. Similarly, the marine copepod

Acartia tonsa feeding on phytoplankton (*Rhodomonas baltica*) showed an increase in foraging effort (measured by the beating of cephalic appendages) as prey levels increased, ostensibly as a result of switching from a passive to an active foraging mode as the energetic benefit of active foraging increased with prey abundance (Kiørboe et al., 2018). Barrios-O'Neill et al. (2014) found that type III functional responses emerged with the addition of predator-free space, such that as prey became denser they became increasingly at risk of encountering predators. This would be consistent with either an increase in detectability (d) or an increase in searching velocity of the prey (V_r). Similarly, sea stars (*Asterias vulgaris*) foraging on sea scallops (*Placopecten magellanicus*) had lower encounter rates and longer handling times at low densities when foraging on heterogeneous substrates than on homogeneous substrate, generating a more type III shape on heterogeneous substrates (Wong et al., 2006). This may have been due to the ability of sea scallops to access refugia in the heterogeneous substrates. Hammill et al. (2010) found a switch from a type II to a type III functional response for *Stenostomum* flatworms foraging on *Paramecium* after the *Paramecium* had been pre-exposed to predation. The exposed *Paramecium* became wider and moved more slowly, which could reduce attack successes or encounter rates, respectively. But the type III response suggests that, with increasing abundance, the *Stenostomum* increased their foraging on *Paramecium*, which might have happened if the *Stenostomum* were able to adjust to the new shape and swimming behavior of the *Paramecium* with the increased practice afforded by greater contacts as prey levels increased (Bergelson, 1985; Hammill et al., 2010). This would be consistent with an increase in probability of success (p_s) with increasing prey levels owing to some form of learning or practice. Finally, Daugaard et al. (2019) found that type III functional responses occurred for the ciliate *Spathidium* sp. foraging on *Dexiostoma campylum* at colder temperatures but reverted to type II at warmer temperatures. This finding suggests that temperature can alter the relationship between underlying mechanisms of space clearance rate and prey density, greatly increasing the challenge of both identifying and understanding the origin of type III functional responses. As a result of these diverse results, it currently seems that there is no consistent conceptual or mathematical underpinning of the type III functional response.

Type III functional responses also may emerge at the population scale if the abundance of prey is correlated in space and/or time with environmental factors that are related to foraging, such as visibility, temperature, or prey searching velocities. For example, Morozov (2010) suggested that vertical migration of zooplankton causes them to forage in different habitats with different local type II functional responses. If the steeper functional response

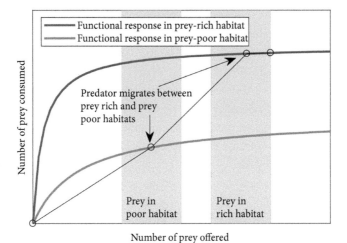

Figure 9.1 continued:

- Legend: Functional response in prey-rich habitat; Functional response in prey-poor habitat
- Y-axis: Number of prey consumed
- Label: Predator migrates between prey rich and prey poor habitats
- Labels: Prey in poor habitat; Prey in rich habitat
- X-axis: Number of prey offered

Figure 9.1 Type III functional responses can emerge at the population level if prey levels across habitats are correlated with functional response parameters. In the case proposed by Morozov (2010), a predator migrates daily between a prey-rich patch and a prey-poor patch. If the functional response is higher in the prey-rich patch, possibly because of different environmental conditions, then the observed foraging rates across prey levels could be sigmoidal. Such an "emergent" type III functional response (the thin line connecting the patch-specific curves) is not explained by changes in the parameters with prey levels but by the movement between patches.

occurs in the habitat with higher prey density, then the overall, average functional response would emerge as type III (Figure 9.1). This outcome could be important for modeling populations that have variation in local functional responses but are being pooled across habitats. However, calling such an emergent functional response type III is somewhat misleading. It would be better to make parameters of the functional response functions of habitat or environmental conditions, so that the functional response itself can respond to local conditions, in whatever way conditions might change in time and space.

Type III functional responses may make most sense in the context of a multi-species functional response (MSFR), when a predator has the ability to switch *off* from a rare prey and focus on something more abundant. For example, an MSFR for hen harriers suggested that they would switch off red grouse toward meadow pipits (*Anthus pratensis*) and field vole (*Microtus agrestis*) when grouse were rare, but would not switch off pipits and voles when they were rare (Smout et al., 2010). Similarly, rough-legged buzzards (*Buteo lagopus*) showed a type II functional response for Norwegian lemmings (*Lemmus lemmus*) and a type III functional response for gray-sided voles (*Myodes rufocanus*), indicating switching off from voles when rare but continued exploitation of lemmings even when rare (Hellström et al., 2014). Indeed, if predators showed type III functional responses for all prey types, then they would switch off everything when rare and starve (Morozov and Petrovskii, 2013). This unfavorable outcome suggests that type III functional responses in nature should only show up for prey that are not worth pursuing when rare even when the predator is hungry. This idea also suggests that MSFRs should always contain at least one prey type that remains type II under any level of alternative prey.

The type III functional response poses problems for optimal foraging theory (OFT) (Chapter 7). Equation (7.7) clearly indicates that the abundance of a less rewarding prey is not a factor in determining the decision to eat it. Only when the abundance of a more rewarding prey declines enough should a predator decide to include the less rewarding prey in the diet. At first pass this suggests that type III predators are behaving in a way that is contradictory to OFT. However, since OFT focuses on the decision to attack, it would only be contradictory if a predator's p_a increased with prey density, as could be the case for dragonfly foraging (Bergelson, 1985). It would be possible, then, for a predator foraging according to OFT to display a type III functional response if space clearance rate changed with prey abundance through other mechanisms, such as probability of success, detectability, or velocities.

9.2 Concerns about the standard type III model

The standard type III functional response (equation (2.8)) is in wide use and is the starting point for assessing the presence of a sigmoidal shape in most studies. In this model, space clearance rate is an increasing function of prey density. In the traditional view, space clearance rate is a linear function of prey abundance, $a = bR$. As a result, the functional response model becomes:

$$f_{pc} = \frac{bR^2}{1 + bR^2 h} \tag{9.1}$$

The squared term is still taken as the default version of type III occasionally, but the flexible versions of this model add an exponent q to R (Chapter 2), to give $a = bR^q$ and:

$$f_{pc} = \frac{bR^{q+1}}{1 + bR^{q+1} h} \tag{9.2}$$

With this model, the q can be estimated along with the other parameters (see Chapter 10).

Although this model can be well fit to some data sets, the linear or exponent versions of the dependence of a on R are somewhat problematic for three reasons. First, given the breakdown of space clearance rate, we know that none of the mechanisms generating space clearance rate can increase in an open-ended way with prey levels. Both probability of attack and success are capped at one, detection distance must have a maximum reflecting the physical ability to sense prey, and neither predator nor prey are able to increase searching velocities indefinitely with prey level. Thus, although some mechanisms might increase linearly over some prey range, it is not biologically reasonable for q to be larger than one because this implies that space clearance rate continually accelerates with prey levels. Exponents less than one are somewhat more plausible, since this implies that space clearance rate increases with prey levels at a decelerating rate, but this expression still does not have the kind of limitations that should arise as predators hit up against the mechanistic limitations of the foraging process.

Second, the parameter b is not biologically equivalent to space clearance rate. Recall that space clearance rate is the slope of the type II functional response as it passes through the origin (Figure 2.2A). The parameter b does not have that interpretation, and it is not clear that there is any way to identify a realized level of space clearance rate from a type III model that is biologically equivalent to this primary definition. As a result, one cannot compare a and b across type II and type III fits and gain any understanding about what is different between the two predator–prey interactions. Indeed, to fit a type III functional response, the value of b must be much less than the value of a would be if the function were fit as type II. To illustrate, I drew a standard type II functional response, with $a = 0.2$ and $h = 1$ (Figure 9.2). I then drew a type III functional response with $q = 0.3$, which required setting b to the lower value of 0.07 to achieve a roughly similar shape. Using higher values of q requires still lower values of b to give a similar shape.

Third, the realized space clearance rate for this functional response starts low, as intended, but it increases rapidly with prey levels (Figure 9.2A). In fact, the realized space clearance rate ($a = bR^q$) may eventually reach a very unrealistic level at higher prey levels. In the example above, the realized space clearance rate exceeds the original a of 0.2 around $R = 30$. This outcome results in a problematic and generally unnoticed phenomenon. If we set the handling time of our type II and type III functional response to zero (reducing the former to a type I), we see that the predicted foraging rate from just the bR^q function crosses the type I curve and continues to accelerate (Figure 9.2B).

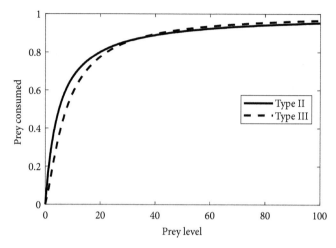

Figure 9.2 Similar shapes of functional responses can be generated with a standard type II and standard type III models if a $q > 0$ is paired with a b lower than the space clearance rate in the type II case. Here, the type II functional response is drawn with $a = 0.2$ and $h = 1$, and the type III response is drawn with $q = 0.3$ and $b = 0.07$ and the same handling time.

This outcome suggests that the standard type III function generates predator–prey interactions that exceed what is expected by mass action. This makes no sense. The mass action starting point is just that, a starting point of good conditions, while the type II and other modifications reduce foraging rate as other processes are taken into account. The flaw in this mechanism is hidden in practice because handling time suppresses the foraging rate as prey levels increase, preventing the acceleration of foraging to these unrealistic levels.

9.3 An alternative type III model

My general sense of a type III functional response is that predator–prey interactions are suppressed at low prey abundances but converge on what might happen in a type II scenario as prey levels increase. Thus, it would make sense to speculate that our standard space clearance rate, a, is a maximum value that describes the interaction without the complications of, say, prey refuges, and that as prey levels decline, the realized space clearance rate declines toward zero reflecting some interruption of the foraging process (reduced velocity, detection, probability of attack, and success). This could be captured by a variety of functions that generate an asymptotic a, such as $\frac{aR}{k_R + R}$, where k_R is a half-saturation constant, or the level of R where the

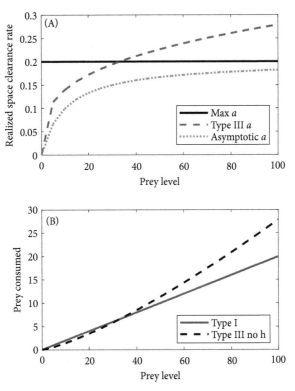

Figure 9.3 Behavior of the realized space clearance rate for a standard type III functional response. (**A**) In type II responses, space clearance rate is constant (max *a*), while in a type III response, space clearance rate is an increasing function of prey level (type III *a*). Alternatively, space clearance rate could be thought to approach a maximum (asymptotic *a*). (**B**) As a result of the realized space clearance rate continually increasing, the model has the potential to predict unreasonably high foraging rates. Setting $h = 0$, the increasing space clearance rate generates foraging rates that exceed what is possible given mass action.

realized space clearance rate reaches half of the maximum value (Figure 9.2B). Inserting this expression into the standard type II functional response equation gives:

$$f_{pc} = \frac{\left(\dfrac{aR}{k_R+R}\right) R}{1 + \left(\dfrac{aR}{k_R+R}\right) hR} \tag{9.3}$$

This expression generates a low realized space clearance rate at low prey levels that then approaches the standard space clearance rate as R increases, causing a sigmoidal shape (Figure 9.3A). And, when $k_R \to 0$, the function reduces to a type II functional response, with a having the same biological meaning and units as it does in the type II model. Furthermore, this "asymptotic a" model recognizes that space clearance rate has limits and cannot increase indefinitely with prey levels, and it cannot by definition exceed the mass action expectation. Following the derivation of the Roger's Random Predator (RRP) equation (Box 9.1), equation (9.3) can be integrated to yield:

$$R_e = R_0\left(1 - e^{\left[a(hR_e - t) - k_R\left(\frac{1}{R_0} - \frac{1}{R_0 - R_e}\right)\right]}\right) \tag{9.4}$$

Box 9.1. Derivation of the asymptotic a equation.

Refollowing the steps of the RRP equation, we start with a differential equation that describes the rate of change in prey abundance (R) through time (t) owing to predation:

$$\frac{dR}{dt} = -\frac{\left(\frac{aR}{k_R + R}\right)R}{1 + \left(\frac{aR}{k_R + R}\right)hR}$$

If we integrate the equation over the course of an experiment of duration t, we will get the cumulative loss of prey from the amount of prey offered at the beginning, R_0, to the number of prey items left at the end of the experiment, R_t. First we arrange the equation to aggregate R to the left-hand side and time to the right-hand side:

$$\frac{\left(1 + \left(\frac{aR}{k_R + R}\right)hR\right)(k_R + R)}{R^2}dR = -adt$$

Expand the left-hand side using FOIL (first, outer, inner, last) and do some factoring out:

$$\frac{k_R}{R^2} + \frac{1}{R^2} + ahdR = -adt$$

Now take the integrals from R_0 to R_t and from zero to t on the left-hand side and right-hand side, respectively:

$$\int_{R_0}^{R_t}\left[\frac{k_R}{R^2} + \frac{1}{R^2} + ah\right]dR = \int_0^t -adt$$

Integration yields:

$$k_R \left(\frac{1}{R_o} - \frac{1}{R_t} \right) + \ln R_t - \ln R_o + ahR_t - ahR_o = -at$$

which we can rearrange as:

$$\ln \frac{R_t}{R_o} = -at - ahR_t + ahR_o - k_R \left(\frac{1}{R_o} - \frac{1}{R_t} \right)$$

and then further to:

$$\ln \frac{R_t}{R_o} = a(-t - hR_t + hR_o) - k_R \left(\frac{1}{R_o} - \frac{1}{R_t} \right)$$

Finally, exponentiating both sides followed by moving the R_0 to the right-hand side gives:

$$R_t = R_o e^{\left[a(h(R_o - R_t) - t) - k_R \left(\frac{1}{R_o} - \frac{1}{R_t} \right) \right]}$$

Following the same tricks we used with the RRP equation, we arrive at equation (9.4):

$$R_e = R_o \left(1 - e^{\left[a(hR_e - t) - k_R \left(\frac{1}{R_o} - \frac{1}{R_o - R_e} \right) \right]} \right)$$

It appears that following Bolker's derivation does not lead to a useful closed-form version of this model. Nonetheless, equation (9.4) provides a more biologically meaningful view of type III functional responses because the new parameter, k_R, indicates a specific feature of the curve; that is, at what point along the curve does realized space clearance rate drop to half its maximum value. This point could arise through a variety of changes, as indicated by our space clearance rate expression. For example, predators could reduce their probability of attack p_a by half at some prey density, possibly owing to the presence of other, more profitable prey. Likewise, prey refuges could reduce the detectability of prey by half at k_R. Nonetheless, k_R is still phenomenological, as it describes the outcome of some effect rather than the mechanism itself.

To illustrate, I fit the asymptotic a model to data of the marine copepod *Paracartia grani* foraging on the algae *Isochrysis galbana* (Helenius and Saiz, 2017) (Figure 9.4). In this study, the original authors found this functional response to be type III using the standard model. How would it look fitted

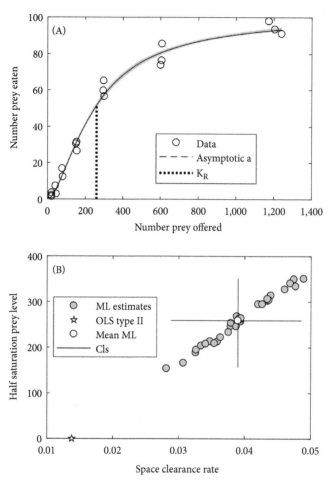

Figure 9.4 Fit of the asymptotic *a* model to foraging data of the copepod *Paracartia grani* foraging on the algae *Isochrysis galbana*. Data from Helenius and Saiz (2017). (**A**) Functional response with the median fit (dark black line) and the bootstrapped fit (narrow cluster of gray lines around the black line). The dotted vertical line marks the estimated k_R. Note that the data from the original were rescaled to the number of prey in 0.1 mL. (**B**) Estimates of space clearance rate (units are 0.1 mL per predator per day) and k_R (prey per 0.1 mL). The star marks the type II estimate (with $k_R = 0$). Note the clear indication of structural correlation between space clearance rate and k_R.

with the asymptotic *a* model? The challenge here is that without a closed-form solution, and with prey depletion, we need to somehow fit equation (9.4) to the data. To do this, I used a simple maximum likelihood approach, searching parameter space to find sets of space clearance rates, handling times, and half-saturation constants that produced predicted foraging rates across all the prey levels that were most likely given the data. I did this using a Monte

Carlo Markov Chain (MCMC), where from some initial guess (here from an initial type II fit), I iteratively moved through parameter space looking for improvement. I also used a bootstrapping approach to get confidence intervals. Bootstrapping is just a way of getting different versions of the data set you might have gotten, and it is done by randomly sampling data with replacement to generate alternative datasets. Thus, I found the most likely parameter set for each bootstrapped data set, and the confidence intervals on the estimates come from the percentiles of the bootstrapped parameter distributions.

For the copepod data, the asymptotic a model fit quite well (Figure 9.4A). The data were very well distributed along the prey levels, which helped resolve the shape. Comparing the two models using the Akaike Information Criterion (AIC; which is a way of finding out whether the data support adding another parameter to the model, in this case k_R), the asymptotic a model had considerably more support (ΔAIC $\gg$ 2, which indicates that adding k_R is justified by the data). The median k_R was 228 prey, marked on the figure with the vertical dotted line. In comparison with a standard type II fit to the data, the maximum likelihood approach suggested that the space clearance rate was higher (Figure 9.4B), which is supported by the fact that the confidence intervals on the asymptotic a estimates do not include the type II estimate. It is also worth noting that there is a strong signal of structural correlation between space clearance rate and k_R, similar to the kind of link between space clearance rate and q in the standard model (Figure 9.4B). Combinations of higher space clearance rate and k_R tended to produce similar fits, suggesting some difficulty cleanly identifying these parameters.

10
Statistical Issues in the Estimation of Functional Responses

In this chapter I cover some key issues in fitting functional response models to data and determining the values of parameters. Because some of these issues have been covered elsewhere, here I focus on the nature of foraging trial data and why noise, stochasticity, and individual variation pose particular challenges for understanding functional responses.

10.1 Curve fitting

The basic experimental and statistical approaches to estimating functional responses emerged long ago. Standard functional response experiments involve replicate foraging trials, with predators, typically used only once, allowed to forage for a set amount of time (or less commonly and usually only for experiments with fish, for a certain number of individuals). Yet we are still grappling with how to handle these data, how to estimate parameters robustly, and how to include additional processes and factors in our models that may play a role in biasing estimates or that may capture variation in foraging behavior.

Given a typical foraging experiment, most workers proceed to fit one or more functional response models to the data using some standard statistical approach. "Fitting" amounts to nothing more than finding the curve that looks most like the data given some criteria. Since the beginning of functional response research, researchers have used a wide variety of techniques to fit these curves to data and thereby estimate functional response parameters. Over the years, researchers have introduced, compared, and evaluated numerous methods for fitting these curves, and many of the approaches are still in use (Bolker, 2011; Fan and Petitt, 1994; Houck and Strauss, 1985; Juliano, 2001; Pritchard et al., 2017; Rosenbaum and Rall, 2018; Trexler et al., 1988; Uszko et al., 2020; Williams and Juliano, 1996). There are three main frameworks for finding curves that look like the data, and all three

Predator Ecology: Evolutionary Ecology of the Functional Response. John P. DeLong, Oxford University Press.

are still in use: non-linear least squares regression (Uszko et al., 2020), maximum likelihood (Rosenbaum and Rall, 2018), and Bayesian techniques (Smout et al., 2010). These different methods may work better in some scenarios than others (e.g., least squares may be challenging to use for multi-species functional responses (MSFRs)), but it appears that all of them can identify parameters correctly, given known parameters in simulated data sets. Thus, the choice of technique sometimes may reflect personal preference or past experience. More critical at this point is probably recognizing the many sources of error and bias in the data and in the selection of models to fit.

One major issue in functional response analysis has revolved around problems with the data. In the 1970s, there was a realization that depleting prey levels during the course of a foraging trial represented a violation of the assumptions of the standard type II functional response. A solution to that was given by the Rogers Random Predator (RRP) equation (Rogers, 1972; Royama, 1971), which integrated the disc equation over the course of a foraging trial to provide an explicit solution to the depletion problem (Box 2.3). Then, however, workers dealt with fitting this equation to data. Since it did not have a closed-form solution, it had to be fit with numerical solvers. These issues were dealt with in an approximate way using a variety of transformations (Houck and Strauss, 1985; Williams and Juliano, 1996), but Bolker's derivation of the Lambert Random Predator (LRP) equation solved that problem. Now, for a type II functional response at least, it is clear that we can estimate parameters reliably using a variety of fitting approaches (Rosenbaum and Rall, 2018; Uszko et al., 2020).

However, issues remain even for type II functional responses. In particular, the variance in foraging rates among replicates typically increases at higher prey densities (Casas and Hulliger, 1994; Uiterwaal et al., 2018). This heteroscedasticity can cause fitted parameters to be biased. Uszko et al. (2020) suggested log-transforming the y axis to handle this problem, as this can improve both accuracy and precision of recovering original parameters. This brings up the reality that many foraging experiments are incredibly noisy, especially at higher prey densities. Across replicate trials, the number of prey consumed can vary widely, suggesting a massive amount of unaccounted-for processes, from stochasticity to individual variation in foraging ability or motivation to forage. In addition, small sample sizes can introduce bias because of the way sample size is incorporated into some model comparison criteria (e.g., the Akaike Information Criterion) and because of non-linear averaging issues (Novak and Stouffer, 2020). For example, in a reanalysis of predator-dependent functional responses, Novak and Stouffer (2020) found that smaller sample sizes tended to lead to overestimation of interference levels. Increasing sample sizes can diminish these biases.

The choice of prey levels remains a crucial issue for obtaining reliable parameter estimates. Functional response parameters all have effects on the shape of the functional response curve itself, and generally the parameters affect a particular region of the shape more than others. Referring back to Chapter 2, space clearance rate affects the initial rise, handling time affects the asymptote, and the Hill exponent (or the half-saturation constant in the asymptotic *a* model) affects the sigmoidal aspect of the rise somewhere in the middle. Thus, an issue common to all methods and experiments is that data must be distributed well from low to high densities in order for the shape to be seen and the parameters to be estimated. No statistical approach can capture handling time unless the curve at least begins to saturate. Type III functional responses cannot be detected without sufficient coverage of the densities in the range where the concave-up part of the sigmoid occurs. Thus, recommendations generally include good coverage in the low-density area but increasing distances at high prey levels, such as logarithmic spacing (Uszko et al., 2020). This is particularly important because the functional response can increase in height by increasing space clearance rate or lowering handling time, so without good coverage at both low and high prey levels, different combinations of parameters may equally well describe the data (Novak and Stouffer, 2020; Uszko et al., 2020).

An alternative approach to fitting functional response curves was suggested by Rosenbaum and Rall (2018). Instead of fitting a curve, they suggest solving an ordinary differential equation (ODE) model of the functional response to determine the expected number of prey that would be eaten for a given set of parameters, and then identify the best match between predicted and observed prey consumption across prey levels using maximum likelihood. This approach appears to work as well as any of the others, but can work better if there are unaccounted-for processes occurring in the experiment such as background prey mortality or reproduction. The approach also can allow for prey growth by including a growth term in the ODE model in addition to the functional response term. Neither of those processes is included in the standard functional responses, so ignoring them when they have occurred in a foraging trial forces the fitting routines to ascribe additional mortality or replaced individuals to the functional response parameters, generating bias. Further, whenever there is only an approximate solution to the functional response, this method works better because the ODE solution provides an exact prediction of the number of prey remaining. For functional response models with exact solutions, such as the type II, traditional fitting and the

ODE predicted method work equally well. The one potential issue with this approach is that with additional processes altering the number of prey, there become many more combinations of parameters that can produce the same outcome, yet there is still only one piece of information being used to infer parameters—the number of prey remaining. If functional response experiments have gone on long enough for prey to reproduce, it might be better to fit the ODE to the whole time series so that the shape of the prey decline, which contains more information than merely the last time point, can help distinguish the two processes (e.g., DeLong and Lyon, 2020).

There is now growing recognition that functional responses, generally seen as a description of a pairwise interaction, are really surfaces that may vary within and among predator–prey pairs (Aljetlawi et al., 2004). Given the processes that underlie both space clearance rate and handling time, our view of functional responses should by default be that functional response parameters are not constants but functions of other things. Across predator–prey pairs, it is becoming increasingly clear that functional responses vary along surfaces by body size ratios (Vucic-Pestic et al., 2010), temperature (Uszko et al., 2020), predator density (DeLong and Vasseur, 2011), and many other prey and predator traits (Houck and Strauss, 1985). Although it is clear that we can identify sources of variation in functional response parameters using cross-species analyses (Rall et al., 2012; Uiterwaal and DeLong, 2020), Uszko et al. (2020) more recently suggested that fitting surfaces directly would provide both greater insights into the shape of those surfaces, as well as deal with the potential correlations among functional response parameters. Such an approach will clearly be harder for aggregates of unrelated studies, where numerous factors may differ among experiments, but should be achievable when a surface is envisioned at the start of an experiment, and prey densities and other factors can be distributed appropriately. Fitting surfaces also is a way of recognizing that each individual foraging trial might be influenced by factors other than prey density. The trick is to identify those factors and find a model that appropriately captures the effect of that factor on a functional response parameter (e.g., a power-law effect of body size). Those factors may be generated among individuals, such as individual variation in body condition (Lyon et al., 2018), or be implemented as a treatment, such as temperature.

10.2 Noise and the nature of foraging trial data

Foraging data tend to be noisy, which is both an outcome of the stochastic nature of predator–prey encounters and an issue to overcome for fitting.

To illustrate the importance and meaning of this noise, here I re-estimate the parameters of some functional responses using a variety of approaches and compare their outcomes. As a case study, I will start with the functional responses estimated for Serrano's anole (*Norops serranoi*) foraging on Mexican fruit flies (*Anastrepha ludens*) (Dor et al., 2014) (Figure 10.1). The experiment was conducted as part of an effort to understand the biocontrol potential of the lizards, because they may consume fruit flies (an agricultural pest) they find on the limbs of fruit trees and the stems of yuccas. In this study, the authors captured flies and lizards from the wild and brought them into the lab where they were fed and kept at a common temperature, with the lizards being fed on the same kind of fruit flies that would be used in the experiment. Lizards were kept in 30 × 30 × 30 cm cages that were also used for foraging trials. The researchers offered each lizard between 2 and 60 flies during 6-h foraging trials. Because they had only eight lizards, they used

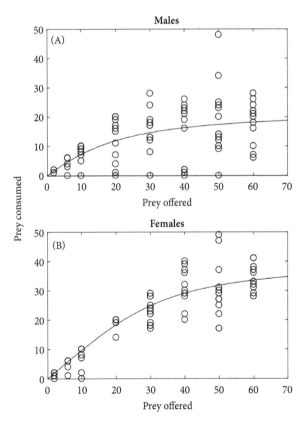

Figure 10.1 Functional responses of male (**A**) and female (**B**) anoles (*Norops serranoi*) foraging on Mexican fruit flies (*Anastrepha ludens*). Data from Dor et al. (2014). Fits are for the Lambert Random Predator equation (type II) using ordinary least squares.

them more than once, but allowed 2-week intervals between repeated trials to increase independence among trials. The researchers did not replenish flies during the experiment, and therefore we need to use either the RRP or the LRP equation to fit the data. I used the original data for this experiment and fit the standard type II LRP equation to the data using ordinary least squares (OLS). With this approach, I estimated the space clearance rate to be 0.22 for males (Figure 10.1A) and 0.67 for females (Figure 10.1B). Because the data are just a total number instead of a density, the units of space clearance rate here are arenas per predator per hour. I estimated the handling time as 0.22 h for males and 0.14 h for females. Visually, it appears that both functional responses are reasonably type II, with a hint of type III for the females. Females appear to have a higher functional response than males, and both data sets show a large amount of noise. Before trying to determine functional response type and whether the curves actually differ between males and females, however, we should explore the nature of data like these.

The first important point regarding functional response data is that they are an amalgamation of foraging trials across different individuals and reflect the stochastic process of predator–prey encounters. Thus, there tends to be a lot of noise in the data along with real among-individual variation that looks (and is treated) like noise. To get a sense for this noise, I simulated the sequence of prey captures in a foraging experiment based on the estimated functional response parameters for male lizards. I did this by estimating the time between captures as $\frac{1}{aR}$, which is the expected interval between captures for a type I functional response (Coblentz and DeLong, 2020). Starting from all the prey offered, I drew a time from an exponential distribution with a mean equal to the expected value, added the handling time, then subtracted one, generating a stepping down of prey levels as they are eaten through time. This gets us a type II functional response where the stochasticity is generated through random encounters rather than random handling times. I ran the simulation for the 6 h of the experiment, but each experiment went to one capture past the 6 h since that interval was determined at the last capture before 6 h was up. I ran 300 simulations at starting prey levels of 10 and 40 flies and then pulled the number of prey remaining at the 6-h mark. In Figure 10.2 it is clear that many different trajectories of prey depletion are possible in a foraging trial governed by the parameters estimated for the lizards. With 10 prey offered and our 1 parameter set, lizards could eat anywhere from 3 to all of the prey in 6 h, with a mode of around 6 prey eaten, which is quite in line with what was seen in the experiment (Figure 10.2A,B). With 40 prey

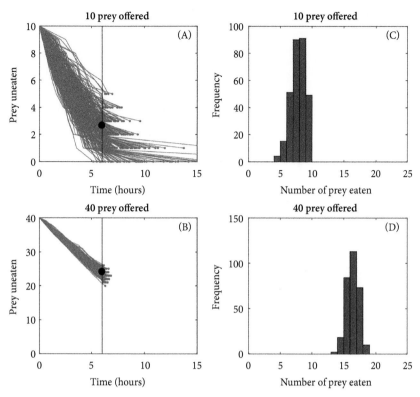

Figure 10.2 Simulated dynamics of individual foraging trials based on the functional response of males in Figure 10.1. Simulations started with either 10 prey (top row) or 40 prey (bottom row); note the different *y* axis ranges. Each of 300 foraging trials are shown in gray in **A** and **C**, while the distribution of number of prey eaten are shown in **C** and **D**. The vertical black line in **A** and **C** marks the 6-h mark when the original experiments ended. The black dot is the average number of prey left after 6 h.

offered, although clearly less stochastic, lizards still showed a wide range of likely foraging levels (Figure 10.2C,D).

These simulations show why functional response data are likely to be very noisy. However, it seems unlikely that this stochasticity can generate all of the noise that we tend to see (Figure 10.3A). Expanding the foraging dynamics simulation to mirror the lizard–fly experiment, with 80 trials spread out over the 8 original prey levels, I created a simulated data set for the whole experiment again given the male's functional response parameters. The simulated data set (filled circles in Figure 10.3A) occupies a noisy swath of space that is much narrower than the actual data (open circles), indicating that the noise in the experiment must also be coming from other sources. I then

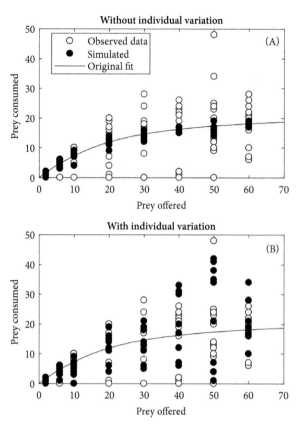

Figure 10.3 Simulated functional response data vs. the original data for male lizards (*Norops serranoi*) foraging on fruit fly (*Anastrepha ludens*). Data from Dor et al. (2014). The observed points are the same as in Figure 10.1A. The simulated points arise from the stochastic simulations of foraging trials shown in Figure 10.2 for all original prey levels. In **A**, there is only stochasticity. In **B**, there is individual variation in the parameters. The point here is that the observed data are noisier than the simulated data without individual variation, suggesting that stochasticity is not sufficient to generate the noise. However, by adding individual variation (standard deviation of 1.2 times the mean parameter values), the data sets that have the right amount of noise.

repeated this exercise but this time I drew a space clearance rate and handling time from a lognormal distribution based on the mean space clearance rate and handling time with a standard deviation of 1.2 times the mean parameter value (Figure 10.3B). Adding this step simulates an experiment with each trial having a different individual predator, and each predator has its own set of parameters. Here it is clear that if there is enough individual variation added then we can recreate the spread of the data. With the standard deviations set at 1.2 times the mean parameter, I could repeatedly get data sets that resembled the original. With less than that, data sets often did not show the same level of spread. This outcome suggests that individual variation in functional response parameters is a potentially important driver of error and noise in foraging trial experiments, as suggested long ago (Houck and Strauss, 1985). Furthermore, individual variation itself could be a factor driving the increased dispersion in the data at higher prey levels (Fan and Petitt, 1994; Uiterwaal et al., 2018; Uszko et al., 2020). However, despite the general visual match, it was rare that these simulations produced the same number of trials that had zero prey consumed. Because trials in which predators do not forage at all are not that uncommon, there may be an important unidentified process influencing the estimates of functional responses.

In the lizard–fly experiment, individuals were used more than once. Yet there were eight individuals that could vary in their body size, age, or other traits that influence space clearance rate components or handling time. These individuals could even vary in motivation to forage during different trials, given variation in their hunger levels through time. Although the traditional functional response experiment provides a group-level outcome, it is becoming increasingly clear that individual variation in functional responses exists and is likely to play an important role in how predator–prey interactions work (Bolnick et al., 2011; Gibert and DeLong, 2017; Rudolf and Rasmussen, 2013; Schröder et al., 2016; Tully et al., 2005). The above simulations hint that there is information about individual variation buried in functional response experiments. Some of this information could be extracted by expanding the functional response into surfaces with traits or other characteristics embedded into the functional response model. For example, in an experiment with the wolf spider *Hogna baltimoriana* foraging on grasshoppers, more variation in the number of prey eaten could be accounted for by incorporating individual energetic state of the spider (body condition, i.e., the ratio of abdomen width to cephalothorax width) into the model as a predictor of space clearance rate (Lyon et al., 2018).

Given that foraging trials typically use individuals only once, and that individuals likely vary in their parameters, fits of functional responses to data may end up underestimating the true average parameters of the population being studied. Both space clearance rate and handling time have non-linear effects on foraging rate in the non-linear models (type II and type III but not type I) (Bolnick et al., 2011) (Figure 10.4). As a result of Jensen's inequality, the average space clearance rate would predict a higher rate of foraging than the actual average foraging rate across individuals (Figure 10.4A). In contrast, the averaging handling time would predict a lower rate of foraging than the average foraging rate across individuals (Figure 10.4B). Since fitting a

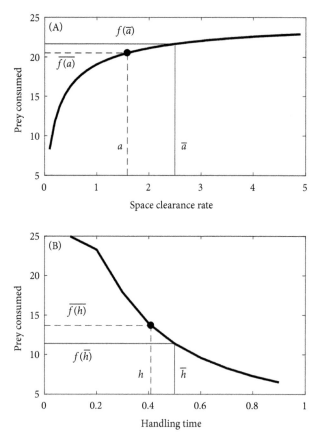

Figure 10.4 Non-linear averaging effect on the estimates of space clearance rate (**A**) and handling time (**B**). Because of the concave down relationship between foraging and space clearance rate, the mean space clearance rate produces a foraging rate higher than the actual, with the reverse occurring for handling time. But curve fitting seeks out the average foraging rate, leading to estimates of both parameters that are lower than the true average in the populations.

functional response model to data is essentially the process of drawing a line through the mean foraging rate at each prey level, the estimated space clearance rate and handling time will be smaller than the actual average parameter in the population (Figure 10.4, filled circles and dashed lines). The upshot is that the true average parameters are higher than estimated whenever individual variation is present, which is nearly all of the time.

These simulations also help bring to light one common issue with functional response experiments that use a set amount of time for each trial in a non-prey-replacement design. It is possible if the time is too long that predators might consume everything in a trial, causing data to occur on the

1:1 line, as did happen occasionally in the lizard–fly experiment (Figure 10.3). This outcome happens in many experiments, especially at low densities. It is an issue because the points along the 1:1 line draw the estimated space clearance rate toward the value of one. A shorter trial would move these points off the 1:1 line and lead to a higher rate owing to the shorter time. In other words, in these trials, there is time included in the experiment that is neither search time nor handling time, and since there is not an "other time" parameter in the model, it will end up being incorporated by either a longer handling time or a smaller space clearance rate in the fit. Therefore, limiting time such that it is short enough to prevent complete prey depletion in the lowest-density trials is good practice. This step also can help to reduce the effect of satiation on within-trial changes in predator motivation to hunt.

10.3 Differences between parameters

A key question about functional responses is how they vary (Chapter 3). Although functional responses might change systematically with body mass, temperature, or arena size, they also may differ between sexes, ages, or ecological factors such as habitat type or whether the predators are invasive or native. In the lizard–fly experiment, the female curve appears higher than the male curve, potentially owing to a higher space clearance rate, a lower handling time, or both. Given the large amount of noise in the data due to stochasticity and individual variation, how can we determine whether the females, on average, actually have different parameters from the males?

There are two straightforward ways of doing this. The first approach is to add an offset term to the model parameters, such that the parameters of two types or treatments can differ from each other by the offset (Juliano, 2001). That is, either parameter can be replaced by $p + \Delta p^* D$, where p is a parameter, D (standing for a "dummy" variable) is either zero or one, and Δp is the offset. Updating the LRP equation (equation (2.17)) to include the offsets for both parameters, we have:

$$R_{\mathrm{e}} = R_0 - \frac{W\left((a + \Delta_a D)(h + \Delta_h D) R_0 e^{-(a+\Delta_a D)(t-(h+\Delta_h D)R_0)}\right)}{(a + \Delta_a D)(h + \Delta_h D)} \qquad (10.1)$$

If you then fit both data sets at once with an extra column for the predictor variable D, the type designated with $D = 0$ has a parameter p, and the type designated as $D = 1$ has a parameter $p+\Delta p$, where one can determine whether the Δp is significantly different from zero using standard approaches. This technique was implemented in the "frair" package for R (Pritchard et al., 2017). I used this technique here for the lizard–fly data set in Figure 10.1 to assess how the functional responses differed between males and females. The estimated offset for space clearance rate was 0.43 (with 95% confidence limits of −0.21 to 1.07), and the estimated offset for handling time was −0.12 (with 95% confidence intervals of −0.22 to −0.029). This result indicates that females (coded with dummy variable one) have a smaller handling time than the males, by 0.12 h, because the confidence limits on the estimate do not include zero. For space clearance rate, although there is a positive estimate of the difference, the confidence limits indicate that it might as easily be zero, indicating we do not have confidence that there is a difference in space clearance rate between males and females.

One downside of this approach is that every pairwise contrast has to be refit to estimate the offsets. Furthermore, in this case, using OLS, the 95% confidence intervals of space clearance rate and handling can sometimes turn out negative, which is not biologically meaningful. A way of dealing with both of these issues is to use bootstrapping. As should be clear from the above simulations (Figure 10.2), any given number of prey consumed that we observe in a foraging trial is just one of many possible numbers, given the inherent stochasticity of the foraging process. We may extend that to say that any given functional response data set is just one of many we could observe generated from the same individuals. Thus, we can estimate parameters from bootstrapped data sets, generating a distribution of possible parameters that could arise from these data (see Chapter 9).

These parameter distributions are also useful for comparing functional responses. We can use all of the pairwise differences among the bootstrapped parameter estimates between groups to get a sense of how often our estimate of the parameter for one group exceeds that of another. In the case of the lizard–fly data, estimates of space clearance rates for males are frequently smaller than those of females. Not all of the pairwise differences are negative, however, so we can use a confidence interval approach to assess the difference. For space clearance rate, 90% of the pairwise differences are negative, which in the traditional sense of statistical significance, would not be significantly different ($P = 0.1$). Some might argue that this difference is "marginally" significant. In contrast, handling time estimates are larger for males than for females 96% of the time ($P = 0.04$), so in this case, the difference is statistically significant. These results align quite well with the offset approach above. Figure 10.5 shows that the estimated offset aligns well with the mode of pairwise differences for both space clearance rate (Figure 10.5A) and handling time (Figure 10.5B). The advantage of the bootstrapped differences approach

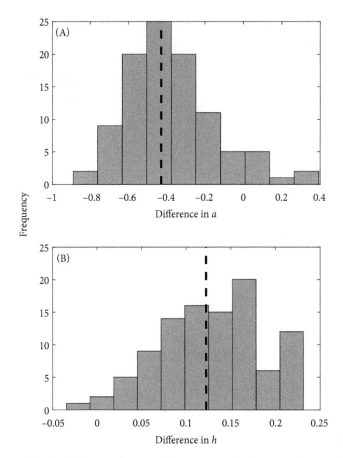

Figure 10.5 Pairwise differences between bootstrapped estimates of space clearance rate (**A**) and handling time (**B**) for male and female lizards (*Norops serranoi*) foraging on fruit fly (*Anastrepha ludens*). Data from Dor et al. (2014). Space clearance rate is not significantly different between males and females, as the distribution crosses the zero mark at the 90th percentile. Females, however, have a lower handling time than males, as the distribution crosses zero at the 96th percentile. The dashed vertical lines show the offset using the dummy variable approach (Juliano, 2001).

is that you get both bootstrapped confidence intervals for parameters and an easy way to compare more than two groups without refitting.

10.4 Type II or type III?

Because of the ecological consequences of switching off from rare prey, a common goal in functional response studies is to determine if the shape of the functional response is type II or type III (Figure 10.6). There are two ways currently used to do this. The first is to evaluate the shape of the proportion of prey consumed with a polynomial (Juliano, 2001; Trexler et al., 1988). This approach benefits from the often-clearer distinction between type II and type III functional responses when portrayed as the proportion of prey eaten rather than the number of prey eaten (Figure 10.6A vs. 10.6B). Although the curves can be very similar over much of their extent, they can be much more clearly distinct in proportion consumed[1] at low prey abundances (Figure 10.6B). To use this technique, one fits a polynomial logistic regression to the proportion of prey consumed to determine whether there is a bend corresponding to a type III. Conformity to the type III shape is determined by a significantly positive linear term and a significantly negative quadratic term (i.e., indicating a downward opening hyperbola; Figure 10.6B), while a type II is revealed by only a significantly negative first-order term (i.e., a generally declining function). After deciding on the type of functional response, one proceeds to fit the appropriate functional response model to the data.

The second way is to fit a type III functional response (such as equation (2.17)) to the data directly and ask whether the exponent on prey abundance is greater than zero. If it is, then the functional response is type III. However, because a type III functional response has more parameters than a type II functional response, it may well fit better without providing a reliable indication that the functional response is really type III. Thus, it may be better to compare the fits using an information criterion that demands an improvement greater than what would be expected by just adding more parameters.

To compare these approaches, I refit the data from a functional response experiment for African clawed frogs (*Xenopus laevis*) foraging on mosquito larvae (*Culex pipiens*) (Thorp et al., 2018) (Figure 10.7A–C). In this experiment, three size classes of predators (small, medium, and large) were used, and the researchers evaluated the type of functional response for the different-sized frogs. I used both the polynomial logistic regression approach and the direct fitting AIC comparison approach to determine whether the functional responses were type II or type III. Using the polynomial approach, for small predators, the cubic term was non-significant, so I removed it and ran just the quadratic model. Here, both the linear and quadratic terms were significantly negative, indicating a type II functional response (Figure 10.7D). For medium predators, the results did not conform to either prediction as the linear term

---

[1] The proportion of prey consumed can also be taken as a measure of risk to the prey, because if the proportion is low, the chance that any given prey individual will be captured is low, implying low risk. This variation in risk may influence prey behavior.

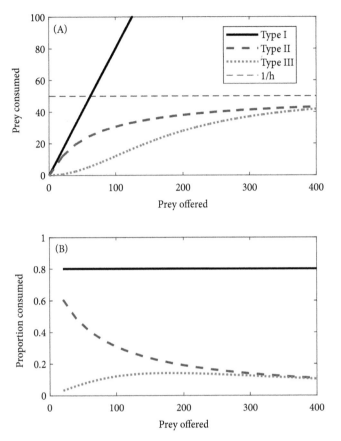

Figure 10.6 Type I, II, and III functional responses by prey consumed (**A**) and by proportion of prey consumed (**B**). Type I functional responses are linear (possibly up to a maximum) and show a constant proportion of prey eaten across prey densities. Type II functional responses are saturating and show a decelerating proportion of prey eaten, as the search time declines with increasing prey abundance. Type III functional responses are sigmoidal, showing a rise and fall in the proportion eaten as the functional response accelerates and then slows down owing to handling time. Because type II and type III functional responses can be difficult to distinguish in the number of prey consumed, an analysis of the proportion can help distinguish the two.

was negative, the quadratic positive, and the cubic negative. That conforms to neither type II nor type III, suggesting that this approach might not be able to distinguish type for this data set (Figure 10.7E). For the large predators, the cubic term was not significant, so I reran this one as a quadratic as well. In this model, the linear term was significantly positive while the quadratic term was significantly negative, conforming to the hump-shaped expectation for a type III functional response (Figure 10.7F).

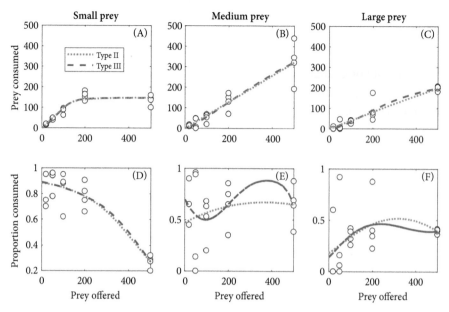

Figure 10.7 Functional responses of small (**A**), medium (**B**), and large (**C**) African clawed frogs (*Xenopus laevis*) foraging on mosquito larvae (*Culex pipiens*). Data from Thorp et al. (2018). The type II fit is a dotted orange line and the type III fit is the dashed blue line. The proportion of prey eaten for small (**A**), medium (**B**), and large (**C**) African clawed frogs, with the quadratic fit (solid black line) and the cubic fit (dashed gray line).

I then fit a type II and type III functional response to each data set, using the LRP approximation equation for at type III functional response with a Hill exponent (Equation 2.18). For small predators, the ΔAIC for type III vs. type II was 0.01, indicating no support for the more complex type III model over the type II. This result is in agreement with the polynomial approach—the small frogs have a type II functional response. For medium predators, the AIC was lower for type II than type III by 1.7, indicating slightly more support for the type II form, which was not identifiable by the polynomial approach. For the large predators, the AIC was again lower for type II than for type III, by 0.62, suggesting no support for the more complex type III model, which contradicts the polynomial result. Thus, the two approaches are not lining up. Is the large frog functional response really type III? Is the polynomial just by chance picking up a quadratic shape because of the vagaries and noise of the data, or is the specific functional form of the type III with an exponent preventing detection of a different sigmoidal shape? A similar problem arises for the lizard–fly dataset, with agreement on type II between the polynomial and model comparison approaches for the males, the polynomial approach suggesting type III for females, but model comparison not supporting the type III model. Although still widely used, the polynomial approach is being gradually displaced by the direct fitting approach estimating the Hill exponent and comparing models with information criteria (Smout et al., 2010; Uiterwaal and DeLong, 2020; Vucic-Pestic et al., 2010). This replacement echoes the gradual replacement of transformation approaches with direct fitting approaches focused on estimating type II functional response parameters (Juliano and Williams, 1985).

11
Challenges for the Future of Functional Response Research

Being recognized for more than 70 years (Solomon, 1949) and estimated more than 2,100 times (Uiterwaal et al., 2018), with numerous analyses of compilations (Englund et al., 2011; Hansen et al., 1997; Jeschke et al., 2004; Kalinoski and DeLong, 2016; Li et al., 2018; Novak and Stouffer, 2020; Rall et al., 2012; Uiterwaal and DeLong, 2018, 2020), it would seem there is a lot we should know about functional responses. Indeed, we know some of the ways in which functional responses vary, how foraging mechanisms combine to determine, to at least some extent, functional response parameters, and how functional responses influence community interactions from biocontrol impacts to invasive predators to food webs. We also know that laboratory-based functional response estimates can be higher than those describing foraging rates in the field (Duijns et al., 2014; Wilhelm et al., 2000). Yet, there remains a considerable amount that we do not know, in particular for field-based functional responses, multi-species functional responses (MSFRs), individual variation, behavioral mechanisms, and the evolution of underlying traits. In this chapter I cover a few key areas for future work.

11.1 MSFRs

As indicated in Chapter 5, much greater attention should be paid to estimating MSFRs. Researchers use MSFRs in community and food web models on a regular basis but have nearly no empirical foundation for that use with the exception of allometric (body mass) based parameterization generated from mostly laboratory studies and a mere handful of field-based MSFRs. Assuming that functional response parameters remain unchanged in the presence of other prey is simple and necessary at this point, but it is likely that these parameters do change most of the time (i.e., are actually surfaces that vary with the presence of other prey types). Therefore, aggregations of

Predator Ecology: Evolutionary Ecology of the Functional Response. John P. DeLong, Oxford University Press.
© John P. DeLong 2021. DOI: 10.1093/oso/9780192895509.003.0011

pairwise functional responses that are not integrated in an MSFR are not likely to make good predictions about foraging rates in communities. After all, would we really expect that no aspect of predator or prey searching behavior, detectability, decisions to attack, or attack success varies with conditions that include alternative prey or the presence of other predators? Indeed, we might even expect foraging behaviors and functional responses to change in the presence of other species that are not in a predator's diet (Kratina et al., 2007), making this issue even more challenging.

It is clear that MSFRs are difficult to estimate experimentally and require more complex fitting approaches, more foraging trials, and greater numbers of individuals, but MSFRs should be a high priority. Bayesian approaches have been employed for the more complex fitting problem of MSFRs, but maximum likelihood should work for these as well. The challenge is that most fitting routines for non-linear models have one dependent variable, but MSFRs have at least two, and possibly many more. Although the Lambert Random Predator (LRP) (Bolker, 2011) equation resolved the analytical problem for prey depletion for one species, the two-species model (Box 5.1) still requires knowing the number of alternative prey consumed, meaning that MSFRs will still require simultaneous solution for multiple dependent variables, unless one just focuses only on a focal prey type with respect to variation in its and other prey levels.

An increase in efforts to estimate field-based MSFRs is encouraging, as the parameter estimates from these analyses reflect the net effects of all the in situ ecological conditions (Hellström et al., 2014; Novak et al., 2017; Smout et al., 2010, 2013). In many cases, however, the density of prey is only known from indices of abundance (e.g., individuals per transect, numbers of scats, etc.), rather than actual densities, meaning that the functional response parameters (mainly space clearance rate because handling time does not have spatial units) themselves would not have units that allow them to be applied to other scenarios or compared among predators or prey (Hellström et al., 2014; Smout et al., 2010). A new approach that estimates foraging rates by determining the proportion of predators detectably processing prey (the "observational" method) may make estimating MSFRs in the field much easier, especially for abundant predators (Novak, 2010; Novak et al., 2017). In this approach, the fact that predators can be found actively processing prey for some amount of time ("detection" time, which again is less than "handling" time) allows estimation of the foraging rate from the proportion of predators handling prey and the detection time. However, even additional effort estimating pairwise functional responses in the field would be welcome (Schenk and Bacher, 2002). For example, O'Donoghue et al. (1998) measured the functional responses of both lynx and coyote (*Canis latrans*) foraging on snowshoe hare over 8 years in the field, revealing differences and similarities between these species. Although their functional response information, gained via tremendous effort, clearly meshed well with observations of the cyclical dynamics of these predator–prey pairs, both lynx and coyote consume other prey, and snowshoe hare has other predators, indicating that even with such a well-studied system we are far from having a thorough depiction of the foraging interactions and their consequences for the ecological communities in which they live.

MSFRs that do not incorporate some form of optimal foraging theory (OFT) and prey selection have the potential to predict suboptimal feeding. That is, if additional prey types become available and the predator consumes them, the predator's overall net energy intake rate (or even just foraging rate) could go down. That is, strictly type II functional responses assume that predators continue to forage on rare prey even when doing so lowers the net energy intake. It is not clear if suboptimal foraging actually occurs in natural systems, but if it does, it suggests that predators may not have the ability to maximize their fitness by adjusting their foraging behavior to maximize energy intake. Of course, predators may be under selection simply to eat more of an abundant prey rather than to switch to a more profitable prey, so optimal foraging is not necessarily the only route to higher predator fitness. Moreover, Kiørboe et al. (2018) showed how the mortality risk and energetic costs of foraging could interact with the net gain from foraging to influence foraging rates and shape functional responses, suggesting that MSFRs might still not be complete without considering the risk of higher trophic level predators on the focal predator.

In contrast, the use of a strictly type III functional response assumes that predators switch off from prey when they are rare, even if doing so leads them to starve. It may take considerable effort estimating MSFRs before we know if suboptimal foraging occurs, or if predators have MSFRs that are flexible and reflect fitness-enhancing decisions, and how to modify functional response models to account for these possibilities (Vallina et al., 2014). In addition to the fitness effects on the predator, adaptive foraging (changing functional response parameters with changing conditions) can stabilize food webs (Kondoh, 2003), but how that foraging adaptation occurs needs more investigation.

11.2 Sources of variation in parameters and constraints

Although there have been several impressive studies evaluating within-population variation in functional responses (Barrios-O'Neill et al., 2015,

2016; Gergs and Ratte, 2009; Miller et al., 1992; Schröder et al., 2016; Spitze, 1985; Tully et al., 2005; Weterings et al., 2015), most of what we know about variation in functional responses comes from across predator–prey pair analyses. This includes the broad effects of predator and prey body size and temperature on functional response parameters (Kalinkat et al., 2011; Li et al., 2018; Rall et al., 2012; Uiterwaal and DeLong, 2018, 2020) and taxonomic differences (Kalinoski and DeLong, 2016; Rall et al., 2011; Uiterwaal and DeLong, 2020). Yet, given the importance of foraging for the fitness of both predator and prey (Chapter 6), the within-population effects of traits on functional response parameters are critical for understanding the evolution of predator–prey interactions, foraging strategies, and prey defenses. Since natural selection acts on phenotypes within populations, the link between traits and functional responses among individuals is a critical part of understanding the evolutionary ecology of the functional response (Tully et al., 2005). It will ultimately be difficult to understand the evolution of foraging or defense traits if we are struggling to connect these traits to their functional consequences for energy gain or survival. These individual-level functional responses are difficult to estimate because of the replication involved (Schröder et al., 2016), but they are a crucial future direction.

Toward that end, a new approach to estimating functional responses has been introduced (Coblentz and DeLong, 2020). In a traditional functional response experiment, each foraging trial yields one data point. This seems a waste, since in some trials numerous predation events have occurred. In this new "time-to-capture" method, the time between kills in a single, or a few, foraging trials can provide enough information to estimate functional responses for individual foragers (see also Tully et al., 2005). This new approach has the potential to expand greatly our ability to estimate functional responses for individuals and understand the phenotypic underpinnings of predator–prey interactions and functional responses. Furthermore, this approach can be used for rare predators and rare prey, which might be too uncommon to find enough individuals for a standard series of trials, to factor in environmental gradients, or to assess individual-based MSFRs.

As well as individual functional responses, additional information may be obtainable by observing the behaviors of predator and prey during foraging trials. Indeed, some recommend studying functional responses by both fitting curves and estimating the parameters directly from behavioral observations (Houck and Strauss, 1985). However, this may not actually be possible. Although handling time is, owing to its name, thought to represent the time spent handling prey, which invokes the idea that the predator is visibly chewing on, holding, or processing the prey, handling time actually is all non-searching time associated with prey capture. That can include behaviors we cannot see, so the prey manipulating time may not line up with the estimated handling time from curve fitting (Schenk and Bacher, 2002; Wong et al., 2006). It also may not be easy to determine when a predator switches from handling to searching again. The same problems arise for observing space clearance rate. This parameter is not just velocity, but the relative velocity, as well as the outcome of predator information gathering and the environment, and their motivation to hunt. However, behavioral observations do provide insight into the source of variation in parameters (Streams, 1994; Wong et al., 2006), even if the observations themselves do not match the estimated parameters. Thus, an increased focus on the behavioral processes and individual traits that influence functional responses could greatly increase our understanding of how functional responses work, why they vary, and how they might be expected to change as ecological conditions change.

One major underexplored area in functional response parameters is the degree to which parameters may be correlated. Because traits that influence foraging processes may influence multiple components of the foraging process, the parameters of the functional response could be mechanistically connected. If so, this would imply areas of parameter space that are not occupied by real systems (Gibert and Wieczynski, 2019). For example, if body size influences the ability of a predator to overpower its prey, it could influence both space clearance rate (through its effect on p_s) and handling time (through its effect on the time to kill or process prey). Similarly, the movement of predators influences both the encounter rate with prey and the encounter rate with other predators. Encounters with competing predators are a potential source of interference competition, so space clearance rate could be positively correlated with interference (DeLong and Vasseur, 2013). Whether other potential correlations could arise, between parameters such as the Hill exponent, wasted time, space clearance rate, and handling time, has not been well explored.

Another key area is the effect of infection status (or "health" in general) in prey or predators on functional response parameters (Seiedy et al., 2012). We know that predators can in some cases detect infections in prey and then choose to avoid them or consume them more (Johnson et al., 2006; Vermont et al., 2016). This suggests that parasites (molecular or organismal) can influence a prey's detectability (d), the probability of attack (p_a), the probability of success (p_s), or even movement behavior. For example, some fish are more likely to consume chytrid-infected *Daphnia pulicaria* than uninfected individuals (Johnson et al., 2006), but is this because they can sense them from chemical signals, prefer to eat infected individuals for nutritional reasons, or because infected *Daphnia* have reduced defenses?

Given how widespread parasites are in food webs (Lafferty et al., 2006, 2008), their role in altering functional responses is probably one of the most important unknowns. For predators, infection can alter the demand for prey, through parasite-induced anorexia, presumably altering searching behavior and the decisions to attack (Hite et al., 2020). For example, crabs (*Eurypanopeus depressus*) infected with a barnacle (*Loxothylacus panopei*) showed a lower functional response on mussels (*Brachidontes exustus*) than uninfected crabs (Toscano et al., 2014). Moreover, variation in age or energy reserves can influence predation risk, decoupling the link between genetically determined traits and the functional response. For example, yellow-legged gulls (*Larus michahellis*) killed by falconry birds were more likely to have low or high muscle condition relative to birds shot by humans (Genovart et al., 2010). Similarly, mountain lions (*Puma concolor*) had higher foraging rates on deer (*Odocoileus hemionus*) when prion infections were high (Miller et al., 2008). In some cases, parasites can alter the potential for predation by manipulating the movement behavior of their hosts (Webster, 2007). Thus, infections in predators and prey are likely to be important in setting functional responses, but there has been little effort to quantify this.

11.3 Functional response models

Numerous functional response models have been proposed, but most researchers still start with a type II functional response and use it if it seems warranted. The mathematical simplicity, frequent good fits to data, and interpretability all support its use, and some studies have claimed that the assumptions of the type II disc equation have indeed been met in experimental trials (Tully et al., 2005). Yet there are at least three key areas that still merit further model development. Given the historical evidence that researchers will prefer to work with simpler models, additional development would benefit from maintaining some mathematical elegance, if it is hoped that the model will get used.

First, the type II functional response assumes that space clearance rate and handling time are constant. This appears to be not completely true— and illogical as well. For foraging trials in which numerous prey are added, it would seem likely that predators get fuller as time goes by, and this might slow down their efforts to find additional food during the trial (Jeschke et al., 2002). Meanwhile, predators provided with few prey remain hungry during the trial. Thus, foraging at high densities may decline because prey are less abundant and because predators become less motivated to forage through time. In one compilation, in fact, it was determined that the longer the foraging trial, the smaller the space clearance rate, suggesting a drop in foraging motivation during experiments (Li et al., 2018). Some additional work to account for reduced motivation to forage both during experiments and across studies and ecological scenarios where predators have had access to different levels of food is necessary. This is essential for dealing with predators in natural contexts, since these predators will not have undergone the typical starvation and hunger standardization procedures used in laboratory experiments. Perhaps some future studies could dispense with hunger standardization and instead measure it and attempt to include it as a predictor in a functional response model by making space clearance rate or handling time a function of a hunger proxy.

Second, why do type III functional responses exist? The standard explanation of "prey switching" does not make much sense and is in any case only one of several possibilities. If a predator is hungry enough to eat numerous individuals of a particular prey at medium density, then those predators, assuming the same hunger level, should be eating them at low density, especially when over the course of an experiment they should remain quite hungry. Most functional response experiments do not seem long enough for predators to undergo any serious learning and update their foraging strategy instantaneously such that a type III functional response would emerge. Rather, it seems more likely that prey are more susceptible at higher densities (higher p_s) owing to prey–prey encounters causing either distraction, displacement from safe locations, or increased overall movement. Although type III functional responses have been invoked as emergent, cross-habitat outcomes, their observation in functional response experiments suggest that there are behavioral mechanisms that remain to be identified that explain the unexpectedly low foraging rates at low densities. One possible source of this low foraging is stochasticity. Because foraging, especially the encounter rate aspect of it, is an inherently stochastic process, there is a reasonable chance that predators capture few, or even no, prey when they are rare, shifting the mean foraging rate down at low densities. Meanwhile, stochasticity would play a smaller role at higher densities, potentially creating outcomes where predators forage in an expected manner at high densities but less than expected at low densities owing to chance alone. Thus, some observed type III functional responses could be artifacts of stochasticity.

Third, predators and prey are not gas molecules. Yet, we assume that they are like gas molecules with the mass-action derivation of the type I functional response and the root sum of squares calculation of relative velocity. Perhaps surprisingly, the models that arise from these assumptions seem to characterize data very well and hold up with respect to their units. It remains to be seen, however, if these assumptions hold up universally, whether it seems fine because most studies are conducted in simplified laboratory ecosystems, or whether it is just the case that they are fine for some species and we decide they are fine for all species. For example, the mass action assumption implies that predator–prey encounters are a linear function of prey density (Figure 3.2). Great tits (*Parus major*), however, showed an accelerating rate of encounters with their winter moth (*Operophtera brumata*) prey, implying a change in search behavior through time (Mols et al., 2004). Ruxton (2005) later speculated that this effect could arise through an adjustment period, wherein the predator increases foraging as it explores the area and determines that risk to itself is low. While both of these works suggested that the observation violated the assumption of mass action, it is also possible that it was actually the assumption of a constant space clearance rate that was violated. Since this experiment occurred within a foraging bout, it could imply instead that time-varying foraging behaviors within experiments are a potential source of error or misunderstanding for functional responses.

Even if the mass action assumption is not exactly correct, perhaps directed movements that are clearly non-random (e.g., the stereotyped spiral swimming patterns of the ciliate *Strombidium sulcatum* (Fenchel and Jonsson, 1988) or the use of roads for movement by wolves (Whittington et al., 2005)) still average out to looking like Brownian motion if the populations are large enough. The way we typically think about mass action is for that of an idealized reaction, but a fuller expression includes exponents on the reactants:

$$f_{pc} = aR^{\alpha}C^{\beta} \tag{10.1}$$

These exponents could be anything and, if not one, could reflect many different types of non-random movement of predators and prey. Thus, an important direction for functional responses (and therefore predator–prey ecology in general) is how particular patterns of movement translate to specific exponents on abundance that might allow us to predict predator–prey encounters. Understanding this translation also may aid with predicting the effects of land-use change and habitat features on predator–prey interactions.

11.4 Linking functional responses from foragers to communities

The vast majority of functional responses have been measured at spatial scales much smaller than the scale the predator and prey are occupying as populations. This difference in scales has the potential to generate parameter estimates that are different from what is actually experienced at the population scale. This could arise from a couple of processes. First, the arena itself. Several studies have shown that the size of an arena influences the estimate of space clearance rate, with the parameter generally getting larger in larger arenas (Bergström and Englund, 2004; Uiterwaal et al., 2019; Uiterwaal and DeLong, 2018), because prey (and sometimes predators) have a tendency to move along arena edges (thigmotaxis), increasing the experienced prey density and thus foraging rates along the edges. Second, most arenas are open and do not contain the kind of habitat complexity experienced by foragers in natural settings. This can make a difference for both the shape and parameter estimates of the functional response. Some studies have made an effort to replicate natural foraging conditions in the arenas themselves (Anderson, 2001; Barrios-O'Neill et al., 2016; Vucic-Pestic et al., 2010; Wasserman et al., 2016), but any sort of difference in habitat between the experimental arena and the natural context in which predators and prey interact could alter the realized parameters. Furthermore, even natural variation in habitat structure can alter functional responses. For example, in the ladybird beetle *Propylea quatuordecimpunctata* foraging on the Russian wheat aphid, *Diuraphis noxia*, functional responses were higher on Indian ricegrass, *Oryzopsis hymenoides*, than on crested wheatgrass, *Agropyron desertorum*, and were more type III in the wheatgrass, possibly owing to a difference in plant structure (Messina and Hanks, 1998). There appears to be no satisfactory way in which we can translate functional response estimates from arenas to nature at this point. On the plus side, this arena simplicity has made it possible to evaluate the effects of body mass, temperature, age, and sex on functional response parameters across predator–prey pairs, but on the negative side, it means that laboratory-based functional response parameters used in population models may not accurately reproduce observed population dynamics in nature.

One possible solution for some predators is to treat arenas as "local," small-scale foraging patches, and then to scale them up by aggregating them. This could involve writing equations for each patch and allowing predators to move among patches in some fashion. Then one could allow them to forage within patches according to an observed patch-level functional response. In this way, the effects of moving among foraging sites would be incorporated as the model scaled from patches to populations. This approach was referred to as the "aggregation" approach in Hunsicker et al. (2011). Alternatively, a more systematic effort to understand how habitat "complexity" influences movement could be made. For example, if most predators and prey show thigmotaxis, then the relevant metric of complexity is the amount of edge within an arena. If this edge density can be measured in the field, then functional response parameters could be corrected or scaled according to the amount of edge where measured and the amount of edge where applied.

A completely different approach is to extract the realized functional response parameters from the dynamics of predator and prey populations. This approach would allow estimates of functional responses at the appropriate scale without needing to find a way to translate results from the arena scale to the population scale. This approach has been undertaken a few times, for example in Chapter 4 for the *Daphnia*–algae and lynx–hare systems, and the approach is referred to as the "calibration" approach in Hunsicker et al. (2011). It also has been used a few times for the *Didinium*–*Paramecium* system. This classic predator–prey system has been a useful laboratory model for nearly a century, and oscillating predator–prey population dynamics can be obtained in short order in microcosms, even though the populations often go extinct (DeLong and Vasseur, 2013; Gause, 1935; Luckinbill, 1973). By fitting a predator–prey ordinary differential equation (ODE) model to these dynamics, several estimates of space clearance rate, handling time, and mutual interference have been obtained across multiple studies (DeLong et al., 2014; DeLong and Lyon, 2020; Jost and Ellner, 2000). Although very small, protist microcosms might represent less of a scale mismatch than with many other predators. Within microcosms, neither the *Didinium* nor the *Paramecium* show much thigmotaxis, utilizing much of the space in the microcosm. Further, the habitat complexity is about the same in a microcosm as in a pond—water with microbes and detritus. Not surprisingly then, the ODE fitting approach and the typical functional response approach provide a similar overall picture of space clearance rate for *Didinium* (Figure 11.1A; data from Hewett, 1980; Jost and Ellner, 2000; DeLong and Vasseur, 2013; DeLong et al., 2014; Li and Montagnes, 2015). However, handling time appears to run a bit longer when estimated with ODE fitting at the population scale (Figure 11.1B). This difference suggests that didinia have additional time costs that interrupt foraging over time in their populations compared with shorter foraging trials. With most other, non-microbial systems, however, much additional work translating functional responses results from experimental arenas to the population scale remains to be done.

The need for population-scale functional responses also extends to the MSFR. For field-based MSFRs, this may generally be what we get. For example, the MSFR for rough-legged buzzards was generated by estimating foraging rates for different breeding sites across space and time, which would be expected to reflect all of the net effects of habitat, alternative

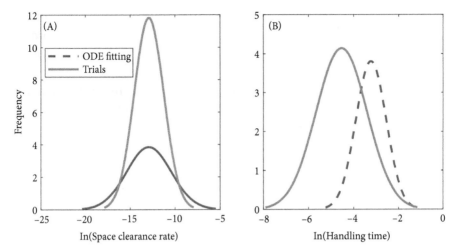

Figure 11.1 Functional response parameters for *Didinium nasutum* foraging on *Paramecium* spp. Space clearance rate is in **A**; handling time in **B**. Different approaches appear to give the same overall range for space clearance rate, but handling times might tend to be a little larger when estimated using ordinary differential equation fitting at the population scale.

prey, other predators, and any other environmental factors (Hellström et al., 2014). Nonetheless, these remain challenging to accomplish, and we generally still need to scale up from the functional responses we have to model communities.

11.5 Accounting for time spent on other activities

Many predators have time budgets that include activities other than searching for and handling prey. These other activities may include sleeping, mating, hiding from other predators, or building nests. If those activities are independent of the number of prey being captured, they would have to be accounted for in the functional response with an additional time budget term. Following the time budget derivation of the functional response from Chapter 2, we can expand our expression for the total time to include some amount of prey-independent time (T_i):

$$T_{tot} = T_s + T_i + aRT_s h \tag{11.1}$$

Dividing the number of prey captured ($F_{pc} = aRT_s$; equation (2.2)) by this total time as we did in Chapter 2 to get the original type II functional response now gets us:

$$\frac{F_{pc}}{T_{tot}} = f_{pc} = \frac{aRT_s}{T_s + T_i + aRT_sh} \tag{11.2}$$

Once again, we can divide the top and bottom by the time spent searching for prey (T_s), yielding this only slightly modified type II model that includes a term for the amount of prey-independent time relative to the search time:

$$f_{pc} = \frac{aR}{1 + \dfrac{T_i}{T_s} + aRh} \tag{11.3}$$

So what then happens to the fitting outcome when you have a predator that has non-zero T_i during the experiment but we fit the data with the standard type II functional response ($f_{pc} = \frac{aR}{1+ahR}$)? To get a handle on this, I will revert to the Michaelis–Menton form of the model by dividing top and bottom by $\frac{1}{ah}$ as we did in equation (2.14):

$$f_{pc} = \frac{aR}{1 + \dfrac{T_i}{T_s} + aRh}\frac{\frac{1}{ah}}{\frac{1}{ah}} = \frac{\frac{1}{h}R}{\frac{1}{ah}\left(1 + \dfrac{T_i}{T_s}\right) + R} = \frac{I_{max}R}{K_R\left(1 + \dfrac{T_i}{T_s}\right) + R} \tag{11.4}$$

This equation suggests that, counterintuitively, estimates of handling time would not be strongly influenced by the presence of unaccounted for prey-independent time, as long as the data allowed a decent estimate of the asymptote itself. That is because the curve still asymptotes at $1/h$ (I_{max}) as R gets large. However, T_i does lower the overall foraging rate by increasing the half-saturation constant (K_R) by a factor of $\left(1 + \dfrac{T_i}{T_s}\right)$. If we are using the standard model without T_i, then space clearance rate has to be smaller (if h is well identified) to achieve the reduced foraging caused by spending time on other activities. This is also apparent because K_R and a are inversely related. What that means in practice is that for any functional response experiment that has gone on long enough for the predators to spend time in non-hunting activities, the space clearance rate may be underestimated. In contrast, the handling time would be unaffected depending on the availability of data at high prey levels, but it also could be underestimated if the data do not allow strong identification of the asymptotic foraging rate. This effect could be what we see for the shorter *Didinium* handling times estimated from foraging trials versus those estimated from time series (Figure 10.1B).

Epilogue

If I have one hope as you reach the end of this book, it is that you agree that our journey to understanding predator–prey interactions is far from over. From the simple fact that the vast majority of predator–prey interactions have never been observed much less quantified as a functional response, to the unsettling reality that we do not yet have the ability to accurately parameterize the functional responses of any complete food web anywhere, should leave us with a sense of purpose. If we had empirically parameterized multi-species functional responses (MSFRs) for food webs, we might have the ability to predict the consequences of climate change, invasive species, or extinctions on the structure and function of ecological communities. At the moment, we can make guesses about how to parameterize such models, but the outcomes of those model simulations are going to be guesses as well. But even an empirically parameterized MSFR is not the end-point—that MSFR needs to be responsive to temperature, predator density, risk, and further evolution of both predator and prey. That is a tall order, but the road to building such a dynamic MSFR is clearly under construction.

I have attempted to show throughout this book that many disparate questions about predators can be linked through the functional response. From foraging behavior, to biocontrol, prey selection, optimal foraging theory, multiple-predator effects, food webs, and evolution of predator–prey interactions, the functional response is the core construct that links them all. I argue that we can use the functional response to ground an integrated science of predator ecology.

The fact that we have measured functional responses thousands of times and that we can see broad patterns such as the effect of body size and temperature should not lead us to think that we know very much about functional responses. It adds up to a lot of information, but even questions as general as whether type III functional responses are common and where they come from are essentially in the same form as they were half a century ago. With emerging new approaches such as the observational method (Novak et al., 2017) and the time-to-capture method (Coblentz and DeLong, 2020), along with more sophisticated approaches for estimating MSFRs (Smout et al., 2010), the real heyday of functional response research is yet to come.

References

Abrams, P.A., 1982. Functional responses of optimal foragers. The American Naturalist 120, 382–390.

Abrams, P.A., 1994. The fallacies of "ratio-dependent" predation. Ecology 75, 1842–1850.

Abrams, P.A., 2000. The evolution of predator–prey interactions: theory and evidence. Annual Review of Ecology and Systematics 31, 79–105.

Abrams, P.A., 2015. Why ratio dependence is (still) a bad model of predation. Biol Rev 90, 794–814. https://doi.org/10.1111/brv.12134

Abrams, P.A., Ginzburg, L.R., 2000. The nature of predation: prey dependent, ratio dependent or neither? Trends in Ecology and Evolution 15, 337–341.

Abrams, P.A., Harada, Y., Matsuda, H., 1993. On the relationship between quantitative genetic and ESS models. Evolution 47, 982–985. https://doi.org/10.2307/2410204

Akcakaya, H.R., Arditi, R., Ginzburg, L.R., 1995. Ratio-dependent predation: an abstraction that works. Ecology 76, 995–1004.

Alexander, M.E., Dick, J.T.A., Weyl, O.L.F., Robinson, T.B., Richardson, D.M., 2014. Existing and emerging high impact invasive species are characterized by higher functional responses than natives. Biology Letters 10, 20130946. https://doi.org/10.1098/rsbl.2013.0946

Alexander, M.E., Kaiser, H., Weyl, O.L.F., Dick, J.T.A., 2015. Habitat simplification increases the impact of a freshwater invasive fish. Environmental Biology of Fishes 98, 477–486. https://doi.org/10.1007/s10641-014-0278-z

Ali, M., Naif, A.A., Huang, D., 2011. Prey consumption and functional response of a phytoseiid predator, *Neoseiulus womersleyi*, feeding on spider mite, *Tetranychus macfarlanei*. Journal of Insect Science 11, 167. https://doi.org/10.1093/jis/11.1.167

Aljetlawi, A.A., Sparrevik, E., Leonardsson, K., 2004. Prey–predator size-dependent functional response: derivation and rescaling to the real world. Journal of Animal Ecology 73, 239–252. https://doi.org/10.1111/j.0021-8790.2004.00800.x

Andersen, R., Saether, B.-E., 1992. Functional response during winter of a herbivore, the moose, in relation to age and size. Ecology 73, 542–550. https://doi.org/10.2307/1940760

Anderson, R.M., Whitfield, P.J., Dobson, A.P., Keymer, A.E., 1978. Concomitant predation and infection processes: an experimental study. Journal of Animal Ecology 47, 891–911. https://doi.org/10.2307/3677

Anderson, T.R., Gentleman, W.C., Sinha, B., 2010. Influence of grazing formulations on the emergent properties of a complex ecosystem model in a global ocean general circulation model. Progress in Oceanography, 3rd GLOBEC OSM: From ecosystem function to ecosystem prediction 87, 201–213. https://doi.org/10.1016/j.pocean.2010.06.003

Anderson, T.W., 2001. Predator responses, prey refuges, and density-dependent mortality of a marine fish. Ecology 82, 245–257.

Aqueel, M.A., Leather, S.R., 2012. Nitrogen fertiliser affects the functional response and prey consumption of *Harmonia axyridis* (Coleoptera: Coccinellidae) feeding on cereal aphids. Annals of Applied Biology 160, 6–15. https://doi.org/10.1111/j.1744-7348.2011.00514.x

Araújo, M.S., Bolnick, D.I., Layman, C.A., 2011. The ecological causes of individual specialisation. Ecology Letters 14, 948–958. https://doi.org/10.1111/j.1461-0248.2011.01662.x

Arditi, R., Perrin, N., Saïah, H., 1991. Functional responses and heterogeneities: an experimental test with cladocerans. Oikos 60, 69–75.

Atkinson, E.C., 1997. Singing for your supper: acoustical luring of avian prey by northern shrikes. Condor 99, 203–206. https://doi.org/10.2307/1370239

Baek, H., 2010. A food chain system with Holling type IV functional response and impulsive perturbations. Computers & Mathematics with Applications 60, 1152–1163. https://doi.org/10.1016/j.camwa.2010.05.039

Barrios-O'Neill, D., Dick, J.T.A., Emmerson, M.C., Ricciardi, A., MacIsaac, H.J., 2015. Predator-free space, functional responses and biological invasions. Functional Ecology 29, 377–384. https://doi.org/10.1111/1365-2435.12347

Barrios-O'Neill, D., Kelly, R., Dick, J.T.A., Ricciardi, A., MacIsaac, H.J., Emmerson, M.C., 2016. On the context-dependent scaling of consumer feeding rates. Ecology Letters 19, 668–678. https://doi.org/10.1111/ele.12605

Baudrot, V., Perasso, A., Fritsch, C., Giraudoux, P., Raoul, F., 2016. The adaptation of generalist predators' diet in a multi-prey context: insights from new functional responses. Ecology 97, 1832–1841.

Beddington, J.R., 1975. Mutual interference between parasites or predators and its effect on searching efficiency. Journal of Animal Ecology 44, 331–340.

Belovsky, G.E., 1978. Diet optimization in a generalist herbivore: the moose. Theoretical Population Biology 14, 105–134. https://doi.org/10.1016/0040-5809(78)90007-2

Bergelson, J.M., 1985. A mechanistic interpretation of prey selection by *Anax junius* larvae (Odonata: Aeschnidae). Ecology 66, 1699–1705. https://doi.org/10.2307/2937365

Bergman, E., 1987. Temperature-dependent differences in foraging ability of two percids, *Perca fluviatilis* and *Gymnocephalus cernuus*. Environmental Biology of Fishes 19, 45–53. https://doi.org/10.1007/BF00002736

Bergström, U., Englund, G., 2004. Spatial scale, heterogeneity and functional responses. Journal of Animal Ecology 73, 487–493. https://doi.org/10.1111/j.0021-8790.2004.00823.x

Berlow, E.L., Dunne, J.A., Martinez, N.D., Stark, P.B., Williams, R.J., Brose, U., 2009. Simple prediction of interaction strengths in complex food webs. PNAS 106, 187–191. https://doi.org/10.1073/pnas.0806823106

Binzer, A., Guill, C., Brose, U., Rall, B.C., 2012. The dynamics of food chains under climate change and nutrient enrichment. Philosophical Transactions of the Royal Society B 367, 2935–2944. https://doi.org/10.1098/rstb.2012.0230

Boit, A., Martinez, N.D., Williams, R.J., Gaedke, U., 2012. Mechanistic theory and modelling of complex food-web dynamics in Lake Constance. Ecology Letters 15, 594–602. https://doi.org/10.1111/j.1461-0248.2012.01777.x

Bolker, B.M., 2011. Ecological Models and Data in R. Princeton University Press.

Bolnick, D.I., Amarasekare, P., Araújo, M.S., Bürger, R., Levine, J.M., Novak, M., Rudolf, V.H.W., Schreiber, S.J., Urban, M.C., Vasseur, D.A., 2011. Why intraspecific trait variation matters in community ecology. Trends in Ecology and Evolution 26, 183–192. https://doi.org/10.1016/j.tree.2011.01.009

Bond, A.B., Kamil, A.C., 1998. Apostatic selection by blue jays produces balanced polymorphism in virtual prey. Nature 395, 594–596. https://doi.org/10.1038/26961

Borrelli, J.J., Allesina, S., Amarasekare, P., Arditi, R., Chase, I., Damuth, J., Holt, R.D., Logofet, D.O., Novak, M., Rohr, R.P., Rossberg, A.G., Spencer, M., Tran, J.K., Ginzburg, L.R., 2015. Selection on stability across ecological scales. Trends in Ecology & Evolution 30, 417–425. https://doi.org/10.1016/j.tree.2015.05.001

Boukal, D.S., Bideault, A., Carreira, B.M., Sentis, A., 2019. Species interactions under climate change: connecting kinetic effects of temperature on individuals to community dynamics. Current Opinion in Insect Science 35, 88–95. https://doi.org/10.1016/j.cois.2019.06.014

Bovy, H.C., Barrios-O'Neill, D., Emmerson, M.C., Aldridge, D.C., Dick, J.T.A., 2015. Predicting the predatory impacts of the "demon shrimp" *Dikerogammarus haemobaphes*, on native and previously introduced species. Biological Invasions 17, 597–607. https://doi.org/10.1007/s10530-014-0751-9

Brose, U., Williams, R.J., Martinez, N.D., 2006. Allometric scaling enhances stability in complex food webs. Ecology Letters 9, 1228–1236. https://doi.org/10.1111/j.1461-0248.2006.00978.x

Burnside, W.R., Erhardt, E.B., Hammond, S.T., Brown, J.H., 2014. Rates of biotic interactions scale predictably with temperature despite variation. Oikos 123, 1449–1456. https://doi.org/10.1111/oik.01199

Butt, A., Talib, R., Khan, M.X., 2019. Effects of insecticides on the functional response of spider *Oxyopes javanus* against aphid *Sitobion avenae*. International Journal of Agriculture and Biology 22, 503–509.

Byström, P., García-Berthou, E., 1999. Density dependent growth and size specific competitive interactions in young fish. Oikos 86, 217–232. https://doi.org/10.2307/3546440

Calder, W.A., 1996. Size, Function, and Life History. Courier Dover Publications.

Carbone, C., Teacher, A., Rowcliffe, J.M., 2007. The costs of carnivory. PLoS Biology 5, e22. https://doi.org/10.1371/journal.pbio.0050022

Carnie, S.K., 1954. Food habits of nesting golden eagles in the coast ranges of California. The Condor 56, 3–12. https://doi.org/10.2307/1364882

Casas, J., Hulliger, B., 1994. Statistical analysis of functional response experiments. Biocontrol Science and Technology 4, 133–145. https://doi.org/10.1080/09583159409355321

Chan, K., Boutin, S., Hossie, T.J., Krebs, C.J., O'Donoghue, M., Murray, D.L., 2017. Improving the assessment of predator functional responses by considering alternate prey and predator interactions. Ecology 98, 1787–1796. https://doi.org/10.1002/ecy.1828

Charnov, E.L., 1976. Optimal foraging: attack strategy of a mantid. The American Naturalist 110, 141–151.

Chesson, J., 1983. The estimation and analysis of preference and its relationship to foraging models. Ecology 64, 1297–1304. https://doi.org/10.2307/1937838

Coblentz, K.E., DeLong, J.P., 2020. Estimating predator functional responses using the times between prey captures. bioRxiv 2020.07.19.208686. https://doi.org/10.1101/2020.07.19.208686

Cock, M.J.W., 1978. The assessment of preference. Journal of Animal Ecology 47, 805–816. https://doi.org/10.2307/3672

Colton, T.F., 1987. Extending functional response models to include a second prey type: an experimental test. Ecology 68, 900–912. https://doi.org/10.2307/1938361

Cortez, M.H., Weitz, J.S., 2014. Coevolution can reverse predator–prey cycles. Proceedings of the National Academy of Science of the United States of America 111, 7486–7491. https://doi.org/10.1073/pnas.1317693111

Coslovsky, M., Richner, H., 2011. Predation risk affects offspring growth via maternal effects. Functional Ecology 25, 878–888. https://doi.org/10.1111/j.1365-2435.2011.01834.x

Crookes, S., DeRoy, E.M., Dick, J.T.A., MacIsaac, H.J., 2019. Comparative functional responses of introduced and native ladybird beetles track ecological impact through predation and competition. Biological Invasions 21, 519–529. https://doi.org/10.1007/s10530-018-1843-8

Crowley, P.H., Martin, E.K., 1989. Functional responses and interference within and between year classes of a dragonfly population. Journal of the North American Benthological Society 8, 211–221.

Curtsdotter, A., Banks, H.T., Banks, J.E., Jonsson, M., Jonsson, T., Laubmeier, A.N., Traugott, M., Bommarco, R., 2019. Ecosystem function in predator–prey food webs—confronting dynamic models with empirical data. Journal of Animal Ecology 88, 196–210. https://doi.org/10.1111/1365-2656.12892

Cuthbert, R.N., Dalu, T., Wasserman, R.J., Callaghan, A., Weyl, O.L.F., Dick, J.T.A., 2019. Using functional responses to quantify notonectid predatory impacts across increasingly complex environments. Acta Oecologica 95, 116–119. https://doi.org/10.1016/j.actao.2018.11.004

Cuthbert, R.N., Dalu, T., Wasserman, R.J., Monaco, C.J., Callaghan, A., Weyl, O.L.F., Dick, J.T.A., 2020a. Assessing multiple predator, diurnal and search area effects on predatory impacts by ephemeral wetland specialist copepods. Aquatic Ecology 54, 181–191. https://doi.org/10.1007/s10452-019-09735-y

Cuthbert, R.N., Wasserman, R.J., Dalu, T., Kaiser, H., Weyl, O.L.F., Dick, J.T.A., Sentis, A., McCoy, M.W., Alexander, M.E., 2020b. Influence of intra- and interspecific variation in predator–prey body size ratios on trophic interaction strengths. Ecology and Evolution 10, 5946–5962. https://doi.org/10.1002/ece3.6332

Czesny, S., Dabrowski, K., Frankiewicz, P., 2001. Foraging patterns of juvenile walleye (*Stizostedion vitreum*) in a system consisting of a single predator and two prey species: testing model predictions. Canadian Journal of Zoology 79, 1394–1400. https://doi.org/10.1139/z01-087

Daugaard, U., Petchey, O.L., Pennekamp, F., 2019. Warming can destabilize predator–prey interactions by shifting the functional response from Type III to Type II. Journal of Animal Ecology 88, 1575–1586. https://doi.org/10.1111/1365-2656.13053

DeAngelis, D.L., Goldstein, R.A., O'Neill, R.V., 1975. A model for trophic interaction. Ecology 56, 881–892.

Dell, A.I., Pawar, S., Savage, V.M., 2014. Temperature dependence of trophic interactions are driven by asymmetry of species responses and foraging strategy. Journal of Animal Ecology 83, 70–84. https://doi.org/10.1111/1365-2656.12081

DeLong, J.P., 2014. The body-size dependence of mutual interference. Biology Letters 10, 20140261. https://doi.org/10.1098/rsbl.2014.0261

DeLong, J.P., 2017. Ecological pleiotropy suppresses the dynamic feedback generated by a rapidly changing trait. The American Naturalist 189, 592–597. https://doi.org/10.1086/691100

DeLong, J.P., 2020. Detecting the signature of body mass evolution in the broad-scale architecture of food webs. The American Naturalist 196, 443–453. https://doi.org/10.1086/710350

DeLong, J.P., Belmaker, J., 2019. Ecological pleiotropy and indirect effects alter the potential for evolutionary rescue. Evolutionary Applications 12, 636–654. https://doi.org/10.1111/eva.12745

DeLong, J.P., Gibert, J.P., 2016. Gillespie eco-evolutionary models (GEMs) reveal the role of heritable trait variation in eco-evolutionary dynamics. Ecology and Evolution 6, 935–945. https://doi.org/10.1002/ece3.1959

DeLong, J.P., Lyon, S., 2020. Temperature alters the shape of predator–prey cycles through effects on underlying mechanisms. PeerJ 8, e9377. https://doi.org/10.7717/peerj.9377

DeLong, J.P., Vasseur, D.A., 2011. Mutual interference is common and mostly intermediate in magnitude. BMC Ecology 11, 1.

DeLong, J.P., Vasseur, D.A., 2012a. A dynamic explanation of size-density scaling in carnivores. Ecology 93, 470–476.

DeLong, J.P., Vasseur, D.A., 2012b. Size-density scaling in protists and the links between consumer–resource interaction parameters. Journal of Animal Ecology 81, 1193–1201. https://doi.org/10.1111/j.1365-2656.2012.02013.x

DeLong, J.P., Vasseur, D.A., 2013. Linked exploitation and interference competition drives the variable behavior of a classic predator–prey system. Oikos 122, 1393–1400. https://doi.org/10.1111/j.1600-0706.2013.00418.x

DeLong, J.P., Hanley, T.C., Vasseur, D.A., 2014. Predator–prey dynamics and the plasticity of predator body size. Functional Ecology 28, 487–493. https://doi.org/10.1111/1365-2435.12199

DeLong, J.P., Gilbert, B., Shurin, J.B., Savage, V.M., Barton, B.T., Clements, C.F., Dell, A.I., Greig, H.S., Harley, C.D.G., Kratina, P., McCann, K.S., Tunney, T.D., Vasseur, D.A., O'Connor, M.I., 2015. The body size dependence of trophic cascades. The American Naturalist 185, 354–366. https://doi.org/10.1086/679735

DeLong, J.P., Hanley, T.C., Gibert, J.P., Puth, L.M., Post, D.M., 2018. Life history traits and functional processes generate multiple pathways to ecological stability. Ecology 99, 5–12. https://doi.org/10.1002/ecy.2070

Denny, M., 2014. Buzz Holling and the functional response. The Bulletin of the Ecological Society of America 95, 200–203. https://doi.org/10.1890/0012-9623-95.3.200

Ding-Xu, L., Juan, T., Zuo-Rui, S., 2007. Functional response of the predator *Scolothrips takahashii* to hawthorn spider mite, *Tetranychus viennensis*: effect of age and temperature. Biocontrol 52, 41–61. https://doi.org/10.1007/s10526-006-9015-7

Doherty, T.S., Glen, A.S., Nimmo, D.G., Ritchie, E.G., Dickman, C.R., 2016. Invasive predators and global biodiversity loss. PNAS 113, 11261–11265. https://doi.org/10.1073/pnas.1602480113

Dor, A., Valle-Mora, J., Rodríguez-Rodríguez, S.E., Liedo, P., 2014. Predation of *Anastrepha ludens* (Diptera: Tephritidae) by *Norops serranoi* (Reptilia: Polychrotidae): functional response and evasion ability. Environmental Entomology 43, 706–715. https://doi.org/10.1603/EN13281

Downing, J.A., 1981. In situ foraging responses of three species of littoral cladocerans. Ecological Monographs 51, 85–104. https://doi.org/10.2307/2937308

Duijns, S., Knot, I.E., Piersma, T., van Gils, J.A., 2014. Field measurements give biased estimates of functional response parameters, but help explain foraging distributions. Journal of Animal Ecology 84, 565–575. https://doi.org/10.1111/1365-2656.12309

Dunn, R.P., Hovel, K.A., 2020. Predator type influences the frequency of functional responses to prey in marine habitats. Biology Letters 16, 20190758. https://doi.org/10.1098/rsbl.2019.0758

Eitzinger, B., Abrego, N., Gravel, D., Huotari, T., Vesterinen, E.J., Roslin, T., 2019. Assessing changes in arthropod predator–prey interactions through DNA-based gut content analysis—variable environment, stable diet. Molecular Ecology 28, 266–280. https://doi.org/10.1111/mec.14872

Endler, J.A., 1986. Defense against predation. In: Feder, M.E., Lauder, G.E. (Eds.), Predator-prey Relationships, Perspectives and Approaches from the Study of Lower Vertebrates. University of Chicago Press, pp. 109–134.

Englund, G., Öhlund, G., Hein, C.L., Diehl, S., 2011. Temperature dependence of the functional response. Ecology Letters 14, 914–921. https://doi.org/10.1111/j.1461-0248.2011.01661.x

Fan, Y., Petitt, F.L., 1994. Parameter estimation of the functional response. Environmental Entomology 23, 785–794. https://doi.org/10.1093/ee/23.4.785

Feldman, C.R., Brodie, E.D., Brodie, E.D., Pfrender, M.E., 2009. The evolutionary origins of beneficial alleles during the repeated adaptation of garter snakes to deadly prey. PNAS 106, 13415–13420. https://doi.org/10.1073/pnas.0901224106

Fenchel, T., 1980a. Suspension feeding in ciliated protozoa: Functional response and particle size selection. Microbial Ecology 6, 1–11. https://doi.org/10.1007/BF02020370

Fenchel, T., 1980b. Suspension feeding in ciliated protozoa: Feeding rates and their ecological significance. Microbial Ecology 6, 13–25. https://doi.org/10.1007/BF02020371

Fenchel, T., Jonsson, P.R., 1988. The functional biology of *Strombidium sulcatum*, a marine oligotrich ciliate (Ciliophora, Oligotrichina). Marine Ecology Progress Series 48, 1–15.

Fox, J.W., 2013. Do you know what a type I functional response is? Are you sure? https://dynamicecology.wordpress.com/2013/06/27/are-you-teaching-type-i-functional-responses-correctly/ (accessed 5.31.16).

Fryxell, J.M., Mosser, A., Sinclair, A.R.E., Packer, C., 2007. Group formation stabilizes predator–prey dynamics. Nature 449, 1041–1043. https://doi.org/10.1038/nature06177

Fussmann, G.F., Blasius, B., 2005. Community response to enrichment is highly sensitive to model structure. Biology Letters 1, 9–12. https://doi.org/10.1098/rsbl.2004.0246

García, L.F., Viera, C., Pekár, S., 2018. Comparison of the capture efficiency, prey processing, and nutrient extraction in a generalist and a specialist spider predator. The Science of Nature 105, 30. https://doi.org/10.1007/s00114-018-1555-z

Gause, G.F., 1935. Experimental demonstration of Volterra's periodic oscillations in the numbers of animals. Journal of Experimental Biology 12, 44–48.

Genovart, M., Negre, N., Tavecchia, G., Bistuer, A., Parpal, L., Oro, D., 2010. The young, the weak and the sick: evidence of natural selection by predation. PLoS ONE 5, e9774. https://doi.org/10.1371/journal.pone.0009774

Gentleman, W., Leising, A., Frost, B., Strom, S., Murray, J., 2003. Functional responses for zooplankton feeding on multiple resources: a review of assumptions and biological dynamics. Deep Sea Research Part II: Topical Studies in Oceanography, The US JGOFS Synthesis and Modeling Project: Phase II 50, 2847–2875. https://doi.org/10.1016/j.dsr2.2003.07.001

Gergs, A., Ratte, H.T., 2009. Predicting functional response and size selectivity of juvenile Notonecta maculata foraging on *Daphnia magna*. Ecological Modelling 220, 3331–3341. https://doi.org/10.1016/j.ecolmodel.2009.08.012

Gibert, J.P., DeLong, J.P., 2017. Phenotypic variation explains food web structural patterns. PNAS 114, 11187–11192. https://doi.org/10.1073/pnas.1703864114

Gibert, J.P., Wieczynski, D.J., 2019. Constraints and variation in food web link-species space. Biology Letters 17, 20210109. https://doi.org/10.1098/rsbl.2021.0109

Gilbert, B., Tunney, T.D., McCann, K.S., DeLong, J.P., Vasseur, D.A., Savage, V., Shurin, J.B., Dell, A.I., Barton, B.T., Harley, C.D.G., Kharouba, H.M., Kratina, P., Blanchard, J.L., Clements, C., Winder, M., Greig, H.S., O'Connor, M.I., 2014. A bioenergetic framework for the temperature dependence of trophic interactions. Ecology Letters 17, 902–914. https://doi.org/10.1111/ele.12307

Ginzburg, L., Colyvan, M., 2004. Ecological Orbits: How Planets Move and Populations Grow. Oxford University Press.

Ginzburg, L.R., Jensen, C.X.J., 2008. From controversy to consensus: the indirect interference functional response. Verhandlungen des Internationalen Verein Limnologie 30, 297–301.

Goss-Custard, J.D., West, A.D., Yates, M.G., Caldow, R.W.G., Stillman, R.A., Bardsley, L., Castilla, J., Castro, M., Dierschke, V., Durell, S.E.A.L.V. dit, Eichhorn, G., Ens, B.J., Exo, K.-M., Udayangani-Fernando, P.U., Ferns, P.N., Hockey, P.A.R., Gill, J.A., Johnstone, I., Kalejta-Summers, B., Masero, J.A., Moreira, F., Nagarajan, R.V., Owens, I.P.F., Pacheco, C., Perez-Hurtado, A., Rogers, D., Scheiffarth, G., Sitters, H., Sutherland, W.J., Triplet, P., Worrall1, D.H., Zharikov, Y., Zwarts, L., Pettifor, R.A., 2006. Intake rates and the functional response in shorebirds (Charadriiformes) eating macro-invertebrates. Biological Reviews 81, 501–529. https://doi.org/10.1017/S1464793106007093

Greene, E., Orsak, L.J., Whitman, D.W., 1987. A tephritid fly mimics the territorial displays of its jumping spider predators. Science 236, 310–312. https://doi.org/10.1126/science.236.4799.310

Gresens, S.E., Cothran, M.L., Thorp, J.H., 1982. The influence of temperature on the functional response of the dragonfly *Celithemis fasciata* (Odonata: Libellulidae). Oecologia 53, 281–284.

Guzman, L.M., Srivastava, D.S., 2019. Prey body mass and richness underlie the persistence of a top predator. Proceedings of the Royal Society B: Biological Sciences 286, 20190622. https://doi.org/10.1098/rspb.2019.0622

Gvoždík, L., Smolinský, R., 2015. Body size, swimming speed, or thermal sensitivity? Predator-imposed selection on amphibian larvae. BMC Evolutionary Biology 15, 238. https://doi.org/10.1186/s12862-015-0522-y

Hammill, E., Petchey, O.L., Anholt, B.R., 2010. Predator functional response changed by induced defenses in prey. The American Naturalist 176, 723–731. https://doi.org/10.1086/657040

Hammill, E., Atwood, T.B., Corvalan, P., Srivastava, D.S., 2015. Behavioural responses to predation may explain shifts in community structure. Freshwater Biology 60, 125–135. https://doi.org/10.1111/fwb.12475

Hansen, J., Bjornsen, P.K., Hansen, B.W., 1997. Zooplankton grazing and growth: scaling within the 2-2,000-um body size range. Limnology and Oceanography 42, 687–704.

Hanski, I., Hansson, L., Henttonen, H., 1991. Specialist predators, generalist predators, and the microtine rodent cycle. Journal of Animal Ecology 60, 353–367. https://doi.org/10.2307/5465

Hartley, A., Shrader, A.M., Chamaillé-Jammes, S., 2019. Can intrinsic foraging efficiency explain dominance status? A test with functional response experiments. Oecologia 189, 105–110. https://doi.org/10.1007/s00442-018-4302-4

Hassanpour, M., Mohaghegh, J., Iranipour, S., Nouri-Ganbalani, G., Enkegaard, A., 2011. Functional response of *Chrysoperla carnea* (Neuroptera: Chrysopidae) to *Helicoverpa armigera* (Lepidoptera: Noctuidae): effect of prey and predator stages. Insect Science 18, 217–224. https://doi.org/10.1111/j.1744-7917.2010.01360.x

Hassell, M.P., 1971. Mutual interference between searching insect parasites. Journal of Animal Ecology 40, 473–486.

Hassell, M.P., Varley, G.C., 1969. New inductive population model for insect parasites and its bearing on biological control. Nature 223, 1133–1137. https://doi.org/10.1038/2231133a0

Hastings, A., 1997. Population Biology: Concepts and Models. Springer-Verlag.

Hector, P., 1985. The diet of the aplomado falcon (*Falco femoralis*) in eastern Mexico. Condor 87, 336–342. https://doi.org/10.2307/1367212

Helenius, L.K., Saiz, E., 2017. Feeding behaviour of the nauplii of the marine calanoid copepod *Paracartia grani* Sars: functional response, prey size spectrum, and effects of the presence of alternative prey. PLoS ONE 12, e0172902. https://doi.org/10.1371/journal.pone.0172902

Hellström, P., Nyström, J., Angerbjörn, A., 2014. Functional responses of the rough-legged buzzard in a multi-prey system. Oecologia 174, 1241–1254. https://doi.org/10.1007/s00442-013-2866-6

Heong, K.L., Bleih, S., Rubia, E.G., 1991. Prey preference of the wolf spider, *Pardosa pseudoannulata* (Boesenberg et Strand). Researches on Population Ecology 33, 179–186. https://doi.org/10.1007/BF02513547

Hewett, S.W., 1980. The effect of prey size on the functional and numerical responses of a protozoan predator to its prey. Ecology 61, 1075–1081.

Hite, J.L., Pfenning, A.C., Cressler, C.E., 2020. Starving the enemy? Feeding behavior shapes host–parasite interactions. Trends in Ecology & Evolution 35, 68–80. https://doi.org/10.1016/j.tree.2019.08.004

Hoddle, M.S., 2003. The effect of prey species and environmental complexity on the functional response of *Franklinothrips orizabensis*: a test of the fractal foraging model. Ecological Entomology 28, 309–318.

Holling, C.S., 1959. The components of predation as revealed by a study of small-mammal predation of the European pine sawfly. The Canadian Entomologist 91, 293–320.

Holling, C.S., 1965. The functional response of predators to prey density and its role in mimicry and population regulation. Memoirs of the Entomological Society of Canada 97, 5–60. https://doi.org/10.4039/entm9745fv

Hossie, T.J., Murray, D.L., 2010. You can't run but you can hide: refuge use in frog tadpoles elicits density-dependent predation by dragonfly larvae. Oecologia 163, 395–404. https://doi.org/10.1007/s00442-010-1568-6

Hossie, T.J., Murray, D.L., 2016. Spatial arrangement of prey affects the shape of ratio-dependent functional response in strongly antagonistic predators. Ecology 97, 834–841. https://doi.org/10.1890/15-1535.1

Houck, M.A., Strauss, R.E., 1985. The comparative study of functional responses: experimental design and statistical interpretation. The Canadian Entomologist 117, 617–629. https://doi.org/10.4039/Ent117617-5

Houde, E.D., Schekter, R.C., 1980. Feeding by marine fish larvae: developmental and functional responses. Environmental Biology of Fishes 5, 315–334. https://doi.org/10.1007/BF00005186

Hunsicker, M.E., Ciannelli, L., Bailey, K.M., Buckel, J.A., Wilson White, J., Link, J.S., Essington, T.E., Gaichas, S., Anderson, T.W., Brodeur, R.D., Chan, K.-S., Chen, K., Englund, G., Frank, K.T., Freitas, V., Hixon, M.A., Hurst, T., Johnson, D.W., Kitchell, J.F., Reese, D., Rose, G.A., Sjodin, H., Sydeman, W.J., van der Veer, H.W., Vollset, K., Zador, S., 2011. Functional responses and scaling in predator–prey interactions of marine fishes: contemporary issues and emerging concepts. Ecology Letters 14, 1288–1299. https://doi.org/10.1111/j.1461-0248.2011.01696.x

Islam, Y., Shah, F.M., Shah, M.A., Musa Khan, M., Rasheed, M.A., Ur Rehman, S., Ali, S., Zhou, X., 2020. Temperature-dependent functional response of *Harmonia axyridis* (Coleoptera: Coccinellidae) on the eggs of *Spodoptera litura* (Lepidoptera: Noctuidae) in laboratory. Insects 11, 583. https://doi.org/10.3390/insects11090583

Jaworski, C.C., Bompard, A., Genies, L., Amiens-Desneux, E., Desneux, N., 2013. Preference and prey switching in a generalist predator attacking local and invasive alien pests. PLoS ONE 8, e82231. https://doi.org/10.1371/journal.pone.0082231

Jeong, H.J., Kim, J.S., Yoo, Y.D., Kim, S.T., Kim, T.H., Park, M.G., Lee, C.H., Seong, K.A., Kang, N.S., Shim, J.H., 2003. Feeding by the heterotrophic dinoflagellate *Oxyrrhis marina* on the red-tide raphidophyte *Heterosigma akashiwo*: a potential biological method to control red tides using mass-cultured grazers. Journal of Eukaryotic Microbiology 50, 274–282.

Jeong, H.J., Yoo, Y.D., Kim, J.S., Seong, K.A., Kang, N.S., Kim, T.H., 2010. Growth, feeding and ecological roles of the mixotrophic and heterotrophic dinoflagellates in marine planktonic food webs. Ocean Science Journal 45, 65–91. https://doi.org/10.1007/s12601-010-0007-2

Jeschke, J.M., 2007. When carnivores are "full and lazy." Oecologia 152, 357–364. https://doi.org/10.1007/s00442-006-0654-2

Jeschke, J.M., Tollrian, R., 2005. Effects of predator confusion on functional responses. Oikos 111, 547–555.

Jeschke, J.M., Tollrian, R., 2000. Density-dependent effects of prey defences. Oecologia 123, 391–396. https://doi.org/10.1007/s004420051026

Jeschke, J.M., Kopp, M., Tollrian, R., 2002. Predator functional responses: discriminating between handling and digesting prey. Ecology Monographs 72, 95–112. https://doi.org/10.2307/3100087

Jeschke, J.M., Kopp, M., Tollrian, R., 2004. Consumer-food systems: why type I functional responses are exclusive to filter feeders. Biological Reviews of the Cambridge Philosophical Society 79, 337–349.

Johnke, J., Baron, M., de Leeuw, M., Kushmaro, A., Jurkevitch, E., Harms, H., Chatzinotas, A., 2017. A generalist protist predator enables coexistence in multitrophic predator–prey systems containing a phage and the bacterial predator *Bdellovibrio*. Frontiers in Ecology and Evolution 5. https://doi.org/10.3389/fevo.2017.00124

Johnson, C.A., Amarasekare, P., Nes, A.E.E.H. van, Day, E.T., 2015. A metric for quantifying the oscillatory tendency of consumer-resource interactions. The American Naturalist 185, 87–99. https://doi.org/10.1086/679279

Johnson, P.T.J., Stanton, D.E., Preu, E.R., Forshay, K.J., Carpenter, S.R., 2006. Dining on disease: how interactions between infection and environment affect predation risk. Ecology 87, 1973–1980. https://doi.org/10.1890/0012-9658(2006)87[1973:dodhib]2.0.co;2

Jost, C., Arditi, R., 2001. From pattern to process: identifying predator?prey models from time-series data. Population Ecology 43, 229–243. https://doi.org/10.1007/s10144-001-8187-3

Jost, C., Ellner, S.P., 2000. Testing for predator dependence in predator-prey dynamics: a non-parametric approach. Proceedings of the Royal Society London B 267, 1611–1620.

Jost, C., Devulder, G., Vucetich, J.A., Peterson, R.O., Arditi, R., 2005. The wolves of Isle Royale display scale-invariant satiation and ratio-dependent predation on moose. Journal of Animal Ecology 74, 809–816.

Joyce, P.W.S., Dickey, J.W.E., Cuthbert, R.N., Dick, J., T.A., Kregting, L., 2019. Using functional responses and prey switching to quantify invasion success of the Pacific oyster, *Crassostrea gigas*. Marine Environmental Research 145, 66–72. https://doi.org/10.1016/j.marenvres.2019.02.010

Juliano, S.A., 2001. Nonlinear curve fitting: predation and functional response curves. In: Cheiner, S.M. and Gurven, J. (Eds.), Design and Analysis of Ecological Experiments. Chapman and Hall, pp. 178–196.

Juliano, S.A., Williams, F.M., 1985. On the evolution of handling time. Evolution 39, 212–215. https://doi.org/10.2307/2408533

Kalinkat, G., Rall, B.C., Vucic-Pestic, O., Brose, U., 2011. The allometry of prey preferences. PLoS ONE 6, e25937. https://doi.org/10.1371/journal.pone.0025937

Kalinoski, R.M., DeLong, J.P., 2016. Beyond body mass: how prey traits improve predictions of functional response parameters. Oecologia 180, 543–550. https://doi.org/10.1007/s00442-015-3487-z

Kiesecker, J.M., Blaustein, A.R., 1997. Population differences in responses of red-legged frogs (*Rana aurora*) to introduced bullfrogs. Ecology 78, 1752–1760. https://doi.org/10.2307/2266098

Kim, J.S., Jeong, H.J., 2004. Feeding by the heterotrophic dinoflagellates *Gyrodinium dominans* and *G. spirale* on the red-tide dinoflagellate *Prorocentrum minimum*. Marine Ecology Progress Series 280, 85–94. https://doi.org/10.3354/meps280085

Kimmance, S.A., Atkinson, D., Montagnes, D.J.S., 2006. Do temperature–food interactions matter? Responses of production and its components in the model heterotrophic flagellate *Oxyrrhis marina*. Aquatic Microbial Ecology 42, 63–73. https://doi.org/10.3354/ame042063

Kiørboe, T., Thomas, M.K., 2020. Heterotrophic eukaryotes show a slow-fast continuum, not a gleaner–exploiter trade-off. PNAS 117, 24893–24899. https://doi.org/10.1073/pnas.2008370117

Kiørboe, T., Saiz, E., Tiselius, P., Andersen, K.H., 2018. Adaptive feeding behavior and functional responses in zooplankton. Limnology and Oceanography 63, 308–321. https://doi.org/10.1002/lno.10632

Klecka, J., Boukal, D.S., 2012. Who eats whom in a pool? A comparative study of prey selectivity by predatory aquatic insects. PLoS ONE 7, e37741. https://doi.org/10.1371/journal.pone.0037741

Koltz, A.M., Classen, A.T., Wright, J.P., 2018. Warming reverses top-down effects of predators on belowground ecosystem function in Arctic tundra. PNAS 115, E7541–E7549. https://doi.org/10.1073/pnas.1808754115

Kondoh, M., 2003. Foraging adaptation and the relationship between food-web complexity and stability. Science 299, 1388–1391. https://doi.org/10.1126/science.1079154

Kopp, M., Tollrian, R., 2003. Trophic size polyphenism in *Lembadion bullinum*: costs and benefits of an inducible offense. Ecology 84, 641–651. https://doi.org/10.1890/0012-9658(2003)084[0641:TSPILB]2.0.CO;2

Korpimäki, E., Norrdahl, K., 1991. Numerical and functional responses of kestrels, short-eared owls, and long-eared owls to vole densities. Ecology 72, 814. https://doi.org/10.2307/1940584

Korpimäki, E., Brown, P.R., Jacob, J., Pech, R.P., 2004. The puzzles of population cycles and outbreaks of small mammals solved? BioScience 54, 1071. https://doi.org/10.1641/0006-3568(2004)054[1071:TPOPCA]2.0.CO;2

Koski, M.L., Johnson, B.M., 2002. Functional response of kokanee salmon (*Oncorhynchus nerka*) to *Daphnia* at different light levels. Canadian Journal of Fisheries and Aquatic Sciences 59, 707–716. https://doi.org/10.1139/f02-045

Kratina, P., Vos, M., Anholt, B.R., 2007. Species diversity modulates predation. Ecology 88, 1917–1923. https://doi.org/10.1890/06-1507.1

Kratina, P., Vos, M., Bateman, A., Anholt, B., 2009. Functional responses modified by predator density. Oecologia 159, 425–433. https://doi.org/10.1007/s00442-008-1225-5

Krebs, C.J., Boonstra, R., Boutin, S., Sinclair, A.R.E., 2001. What drives the 10-year cycle of snowshoe hares? BioScience 51, 25–35. https://doi.org/10.1641/0006-3568(2001)051[0025:WDTYCO]2.0.CO;2

Krebs, J.R., Erichsen, J.T., Webber, M.I., Charnov, E.L., 1977. Optimal prey selection in the great tit (*Parus major*). Animal Behaviour 25, 30–38. https://doi.org/10.1016/0003-3472(77)90064-1

Kreuzinger-Janik, B., Hüttemann, H.B., Traunspurger, W., 2019. Effect of prey size and structural complexity on the functional response in a nematode–nematode system. Scientific Reports 9, 5696. https://doi.org/10.1038/s41598-019-42213-x

Krylov, P.I., 1988. Predation of the freshwater cyclopoid copepod *Megacyclops gigas* on lake zooplankton functional response and prey selection. Archiv für Hydrobiologie 113, 231–250.

Lafferty, K.D., Dobson, A.P., Kuris, A.M., 2006. Parasites dominate food web links. Proceedings of the National Academy of Sciences of the United States of America 103, 11211–11216. https://doi.org/10.1073/pnas.0604755103

Lafferty, K.D., Allesina, S., Arim, M., Briggs, C.J., De Leo, G., Dobson, A.P., Dunne, J.A., Johnson, P.T.J., Kuris, A.M., Marcogliese, D.J., Martinez, N.D., Memmott, J., Marquet, P.A., McLaughlin, J.P., Mordecai, E.A., Pascual, M., Poulin, R., Thieltges, D.W., 2008. Parasites in food webs: the ultimate missing links. Ecology Letters 11, 533–546. https://doi.org/10.1111/j.1461-0248.2008.01174.x

Lam, W., Paynter, Q., Zhang, Z.-Q., 2021. Functional response of *Amblyseius herbicolus* (Acari: Phytoseiidae) on *Sericothrips staphylinus* (Thysanoptera: Thripidae), an ineffective biocontrol agent of gorse. Biological Control 152, 104468. https://doi.org/10.1016/j.biocontrol.2020.104468

Lande, R., 1982. A quantitative genetic theory of Life history evolution. Ecology 63, 607–615. https://doi.org/10.2307/1936778

Lawton, J.H., Beddington, J.R., Bonser, R., 1974. Switching in invertebrate predators. In: Usher, M.B., Williamson, M.H. (Eds.), Ecological Stability. Springer US, pp. 141–158. https://doi.org/10.1007/978-1-4899-6938-5_9

Lee, J.-H., Kang, T.-J., 2004. Functional response of *Harmonia axyridis* (Pallas) (Coleoptera: Coccinellidae) to *Aphis gossypii* Glover (Homoptera: Aphididae) in the laboratory. Biological Control 31, 306–310. https://doi.org/10.1016/j.biocontrol.2004.04.011

Lee, K.P., Simpson, S.J., Clissold, F.J., Brooks, R., Ballard, J.W.O., Taylor, P.W., Soran, N., Raubenheimer, D., 2008. Lifespan and reproduction in *Drosophila*: new insights from nutritional geometry. Proceedings of the National Academy of Sciences of the United States of America 105, 2498–2503. https://doi.org/10.1073/pnas.0710787105

Legett, H.D., Hemingway, C.T., Bernal, X.E., 2020. Prey exploits the auditory illusions of eaves-dropping predators. The American Naturalist 195, 927–933. https://doi.org/10.1086/707719

Lester, P.J., Harmsen, R., 2002. Functional and numerical responses do not always indi-cate the most effective predator for biological control: an analysis of two predators in a two-prey system. Journal of Applied Ecology 39, 455–468. https://doi.org/10.1046/j.1365-2664.2002.00733.x

Levi, T., Wilmers, C.C., 2012. Wolves-coyotes-foxes: a cascade among carnivores. Ecology 93, 921–929.

Li, J., Montagnes, D.J.S., 2015. Restructuring fundamental predator-prey models by recog-nising prey-dependent conversion efficiency and mortality rates. Protist 166, 211–223. https://doi.org/10.1016/j.protis.2015.02.003

Li, Y., Rall, B.C., Kalinkat, G., 2018. Experimental duration and predator satiation levels systematically affect functional response parameters. Oikos 127, 590–598. https://doi.org/10.1111/oik.04479

Lima, S.L., Dill, L.M., 1990. Behavioral decisions made under the risk of predation: a review and prospectus. Canadian Journal of Zoology 68, 619–640. https://doi.org/10.1139/z90-092

Livdahl, T.P., 1979. Evolution of handling time: the functional response of a preda-tor to the density of sympatric and allopatric strains of prey. Evolution 33, 765–768. https://doi.org/10.1111/j.1558-5646.1979.tb04728.x

Líznarová, E., Pekár, S., 2013. Dangerous prey is associated with a type 4 functional response in spiders. Animal Behaviour 85, 1183–1190. https://doi.org/10.1016/j.anbehav.2013.03.004

Luckinbill, L.S., 1973. Coexistence in laboratory populations of *Paramecium aurelia* and its predator *Didinium nasutum*. Ecology 54, 1320–1327. https://doi.org/10.2307/1934194

Lyon, S.R., Sjulin, C.A., Sullivan, K.M., DeLong, J.P., 2018. Condition-dependent foraging in the wolf spider *Hogna baltimoriana*. Food Webs 14, 5–8. https://doi.org/10.1016/j.fooweb.2017.12.003

MacArthur, R.H., Pianka, E.R., 1966. On optimal use of a patchy environment. The American Naturalist 100, 603–609.

MacNulty, D.R., Smith, D.W., Mech, L.D., Eberly, L.E., 2009. Body size and predatory perfor-mance in wolves: is bigger better? Journal of Animal Ecology 78, 532–539. https://doi.org/10.1111/j.1365-2656.2008.01517.x

Maiwald, T., Timmer, J., 2008. Dynamical modeling and multi-experiment fitting with Potter-sWheel. Bioinformatics 24, 2037–2043. https://doi.org/10.1093/bioinformatics/btn350

Mayntz, D., Raubenheimer, D., Salomon, M., Toft, S., Simpson, S.J., 2005. Nutrient-specific foraging in invertebrate predators. Science 307, 111–113. https://doi.org/10.1126/science.1105493

McArdle, B.H., Lawton, J.H., 1979. Effects of prey-size and predator-instar on the predation of *Daphnia* by *Notonecta*. Ecological Entomology 4, 267–275. https://doi.org/10.1111/j.1365-2311.1979.tb00584.x

McCann, K., Hastings, A., Huxel, G.R., 1998. Weak trophic interactions and the balance of nature. Nature 395, 794–798. https://doi.org/10.1038/27427

McCann, K.S., 2011. Food Webs. Princeton University Press.

McCoy, M.W., Bolker, B.M., Warkentin, K.M., Vonesh, J.R., 2011. Predicting predation through prey ontogeny using size-dependent functional response models. The American Naturalist 177, 752–766. https://doi.org/10.1086/659950

McCoy, M.W., Stier, A.C., Osenberg, C.W., 2012. Emergent effects of multiple predators on prey survival: the importance of depletion and the functional response. Ecology Letters 15, 1449–1456. https://doi.org/10.1111/ele.12005

McGhee, K.E., Pintor, L.M., Bell, A.M., Cole, A.E.B.J., Bronstein, E.J.L., 2013. Reciprocal behavioral plasticity and behavioral types during predator–prey interactions. The American Naturalist 182, 704–717. https://doi.org/10.1086/673526

McGill, B.J., Mittelbach, G.C., 2006. An allometric vision and motion model to predict prey encounter rates. Evolutionary Ecology Research 8, 691–701.

McMahon, J.W., Rigler, F.H., 1965. Feeding rate of *Daphnia magna* Straus in different foods labeled with radioactive phosphorus. Limnology and Oceanography 10, 105–113.

McPeek, M.A., 2017. Evolutionary Community Ecology. Princeton University Press.

Messina, F.J., Hanks, J.B., 1998. Host plant alters the shape of the functional response of an aphid predator (Coleoptera: Coccinellidae). Environmental Entomology 27, 1196–1202. https://doi.org/10.1093/ee/27.5.1196

Michalko, R., Pekár, S., Hall, S.R., Bronstein, J.L., 2017. The behavioral type of a top predator drives the short-term dynamic of intraguild predation. The American Naturalist 189, 242–253. https://doi.org/10.1086/690501

Miller, J.R.B., Ament, J.M., Schmitz, O.J., 2014. Fear on the move: predator hunting mode predicts variation in prey mortality and plasticity in prey spatial response. Journal of Animal Ecology 83, 214–222. https://doi.org/10.1111/1365-2656.12111

Miller, M.W., Swanson, H.M., Wolfe, L.L., Quartarone, F.G., Huwer, S.L., Southwick, C.H., Lukacs, P.M., 2008. Lions and prions and deer demise. PLoS ONE 3, e4019. https://doi.org/10.1371/journal.pone.0004019

Miller, T.J., Crowder, L.B., Rice, J.A., Binkowski, F.P., 1992. Body size and the ontogeny of the functional response in fishes. Canadian Journal of Fisheries and Aquatic Sciences 49, 805–812. https://doi.org/10.1139/f92-091

Mols, C.M.M., van Oers, K., Witjes, L.M.A., Lessells, C.M., Drent, P.J., Visser, M.E., 2004. Central assumptions of predator–prey models fail in a semi-natural experimental system. Proceedings of the Royal Society B: Biological Sciences 271, S85–S87.

Monagan, I.V., Morris, J.R., Rabosky, A.R.D., Perfecto, I., Vandermeer, J., 2017. Anolis lizards as biocontrol agents in mainland and island agroecosystems. Ecology and Evolution 7, 2193–2203. https://doi.org/10.1002/ece3.2806

Monteleone, D.M., Duguay, L.E., 1988. Laboratory studies of predation by the ctenophore *Mnemiopsis leidyi* on the early stages in the life history of the bay anchovy, *Anchoa mitchilli*. Journal of Plankton Research 10, 359–372. https://doi.org/10.1093/plankt/10.3.359

Morozov, A., 2010. Emergence of Holling type III zooplankton functional response: Bringing together field evidence and mathematical modelling. Journal of Theoretical Biology 265, 45–54. https://doi.org/10.1016/j.jtbi.2010.04.016

Morozov, A., Petrovskii, S., 2013. Feeding on multiple sources: towards a universal parameterization of the functional response of a generalist predator allowing for switching. PLoS ONE 8, e74586. https://doi.org/10.1371/journal.pone.0074586

Murdoch, W.W., 1969. Switching in general predators: experiments on predator specificity and stability of prey populations. Ecological Monographs 39, 335–354. https://doi.org/10.2307/1942352

Murdoch, W.W., Avery, S., Smyth, M.E.B., 1975. Switching in predatory fish. Ecology 56, 1094–1105. https://doi.org/10.2307/1936149

Nandini, S., Sarma, S.S.S., 1999. Effect of starvation time on the prey capture behaviour, functional response and population growth of *Asplanchna sieboldi* (Rotifera). Freshwater Biology 42, 121–130. https://doi.org/10.1046/j.1365-2427.1999.00467.x

Nilsen, E.B., Linnell, J.D.C., Odden, J., Andersen, R., 2009. Climate, season, and social status modulate the functional response of an efficient stalking predator: the Eurasian lynx. Journal of Animal Ecology 78, 741–751. https://doi.org/10.1111/j.1365-2656.2009.01547.x

Novak, M., 2010. Estimating interaction strengths in nature: experimental support for an observational approach. Ecology 91, 2394–2405. https://doi.org/10.1890/09-0275.1

Novak, M., Stouffer, D.B., 2020. Systematic bias in studies of consumer functional responses. bioRxiv 2020.08.25.263814. https://doi.org/10.1101/2020.08.25.263814

Novak, M., Wolf, C., Coblentz, K.E., Shepard, I.D., 2017. Quantifying predator dependence in the functional response of generalist predators. Ecology Letters 20, 761–769. https://doi.org/10.1111/ele.12777

Oaten, A., Murdoch, W.W., 1975. Switching, functional response, and stability in predator–prey systems. The American Naturalist 109, 299–318.

O'Donoghue, M., Boutin, S., Krebs, C., Zuleta, G., Murray, D.L., Hofer, E.J., 1998. Functional responses of coyotes and lynx to the snowshoe hare cycle. Ecology 79, 1193–1208.

Okuyama, T., 2012. Flexible components of functional responses. Journal of Animal Ecology 81, 185–189. https://doi.org/10.1111/j.1365-2656.2011.01876.x

Oli, M.K., 2003. Population cycles of small rodents are caused by specialist predators: or are they? Trends in Ecology & Evolution 18, 105–107. https://doi.org/10.1016/S0169-5347(03)00005-3

Osenberg, C.W., Mittelbach, G.G., 1989. Effects of body size on the predator-prey interaction between pumpkinseed sunfish and gastropods. Ecological Monographs 59, 405–432. https://doi.org/10.2307/1943074

Otto, S.P., Day, T., 2007. A Biologist's Guide to Mathematical Modeling in Ecology and Evolution. Princeton University Press.

Paine, R.T., 1966. Food web complexity and species diversity. The American Naturalist 100, 65–75. https://doi.org/10.1086/282400

Palkovacs, E.P., Hendry, A.P., 2010. Eco-evolutionary dynamics: intertwining ecological and evolutionary processes in contemporary time. F1000 Biol Reports 2, 1. https://doi.org/10.3410/B2-1

Papanikolaou, N.E., Dervisoglou, S., Fantinou, A., Kypraios, T., Giakoumaki, V., Perdikis, D., 2021. Predator size affects the intensity of mutual interference in a predatory mirid. Ecology and Evolution 11, 1342–1351. https://doi.org/10.1002/ece3.7137

Pawar, S., Dell, A.I., Savage, V.M., 2012. Dimensionality of consumer search space drives trophic interaction strengths. Nature 486, 485–489. https://doi.org/10.1038/nature11131

Pekár, S., Michalko, R., Loverre, P., Líznarová, E., Černecká, Ľ., 2015. Biological control in winter: novel evidence for the importance of generalist predators. Journal of Applied Ecology 52, 270–279. https://doi.org/10.1111/1365-2664.12363

Persson, L., 1986. Temperature-induced shift in foraging ability in two fish species, roach (*Rutilus rutilus*) and perch (*Perca fluviatilis*): implications for coexistence between poikilotherms. Journal of Animal Ecology 55, 829–839. https://doi.org/10.2307/4419

Petchey, O., Beckerman, A., Riede, J., Warren, P., 2008. Size, foraging, and food web structure. Proceedings of the National Academy of Sciences of the United States of America 105, 4191–4196.

Peters, R., 1983. The Ecological Implications of Body Size. Cambridge University Press.

Pierce, G.J., Ollason, J.G., 1987. Eight reasons why optimal foraging theory is a complete waste of time. Oikos 49, 111–118.

Pietrewicz, A.T., Kamil, A.C., 1979. Search image formation in the blue jay (*Cyanocitta cristata*). Science 204, 1332–1333. https://doi.org/10.1126/science.204.4399.1332

Portalier, S.M.J., Fussmann, G.F., Loreau, M., Cherif, M., 2019. The mechanics of predator–prey interactions: first principles of physics predict predator–prey size ratios. Functional Ecology 33, 323–334. https://doi.org/10.1111/1365-2435.13254

Post, D.M., Palkovacs, E.P., 2009. Eco-evolutionary feedbacks in community and ecosystem ecology: interactions between the ecological theatre and the evolutionary play. Philosophical Transactions of the Royal Society B 364, 1629–1640. https://doi.org/10.1098/rstb.2009.0012

Pritchard, D.W., Paterson, R.A., Bovy, H.C., Barrios-O'Neill, D., 2017. frair: an R package for fitting and comparing consumer functional responses. Methods in Ecology and Evolution 8, 1528–1534. https://doi.org/10.1111/2041-210X.12784

Prudic, K.L., Stoehr, A.M., Wasik, B.R., Monteiro, A., 2015. Eyespots deflect predator attack increasing fitness and promoting the evolution of phenotypic plasticity. Proceedings of the Royal Society B 282, 20141531. https://doi.org/10.1098/rspb.2014.1531

Quammen, D., 2004. Monster of God: The Man-Eating Predator in the Jungles of History and the Mind, revised edition. W. W. Norton & Company.

Quinn, T.P., Gende, S.M., Ruggerone, G.T., Rogers, D.E., 2003. Density-dependent predation by brown bears (*Ursus arctos*) on sockeye salmon (*Oncorhynchus nerka*). Canadian Journal of Fisheries and Aquatic Sciences 60, 553–562. https://doi.org/10.1139/f03-045

Rall, B.C., Guill, C., Brose, U., 2008. Food-web connectance and predator interference dampen the paradox of enrichment. Oikos 117, 202–213. https://doi.org/10.1111/j.2007.0030-1299.15491.x

Rall, B.C., Kalinkat, G., Ott, D., Vucic-Pestic, O., Brose, U., 2011. Taxonomic versus allometric constraints on non-linear interaction strengths. Oikos 120, 483–492. https://doi.org/10.1111/j.1600-0706.2010.18860.x

Rall, B.C., Brose, U., Hartvig, M., Kalinkat, G., Schwarzmüller, F., Vucic-Pestic, O., Petchey, O.L., 2012. Universal temperature and body-mass scaling of feeding rates. Philosophical Transactions of the Royal Society B 367, 2923–2934. https://doi.org/10.1098/rstb.2012.0242

Real, L.A., 1977. The kinetics of functional response. The American Naturalist 111, 289–300.

Rendon, D., Taylor, P.W., Wilder, S.M., Whitehouse, M.E.A., 2019. Does prey encounter and nutrient content affect prey selection in wolf spiders inhabiting Bt cotton fields? PLoS ONE 14, e0210296. https://doi.org/10.1371/journal.pone.0210296

Resano-Mayor, J., Real, J., Moleón, M., Sánchez-Zapata, J.A., Palma, L., Hernández-Matías, A., 2016. Diet–demography relationships in a long-lived predator: from territories to populations. Oikos 125, 262–270. https://doi.org/10.1111/oik.02468

Rindorf, A., Gislason, H., 2005. Functional and aggregative response of North Sea whiting. Journal of Experimental Marine Biology and Ecology 324, 1–19. https://doi.org/10.1016/j.jembe.2005.04.013

Ripple, W.J., Beschta, R.L., 2012. Trophic cascades in Yellowstone: The first 15 years after wolf reintroduction. Biological Conservation 145, 205–213. https://doi.org/10.1016/j.biocon.2011.11.005

Roberts, E.C., Wootton, E.C., Davidson, K., Jeong, H.J., Lowe, C.D., Montagnes, D.J.S., 2010. Feeding in the dinoflagellate *Oxyrrhis marina*: linking behaviour with mechanisms. Journal of Plankton Research 33, 603–614. https://doi.org/10.1093/plankt/fbq118

Roemer, G.W., Donlan, C.J., Courchamp, F., 2002. Golden eagles, feral pigs, and insular carnivores: how exotic species turn native predators into prey. PNAS 99, 791–796. https://doi.org/10.1073/pnas.012422499

Rogers, D., 1972. Random search and insect population models. Journal of Animal Ecology 41, 369–383. https://doi.org/10.2307/3474

Rojo, C., Salazar, G., 2010. Why are there so many kinds of planktonic consumers? The answer lies in the allometric diet breadth. Hydrobiologia 653, 91–102. https://doi.org/10.1007/s10750-010-0346-0

Rosenbaum, B., Rall, B.C., 2018. Fitting functional responses: direct parameter estimation by simulating differential equations. Methods in Ecology and Evolution 9, 2076–2090. https://doi.org/10.1111/2041-210X.13039

Rosenzweig, M.L., MacArthur, R.H., 1963. Graphical representation and stability conditions of predator-prey interactions. The American Naturalist 97, 209–223.

Royama, T., 1971. A comparative study of models for predation and parasitism. Researches in Population Ecology 13, 1–91. https://doi.org/10.1007/BF02511547

Rudolf, V.H.W., Rasmussen, N.L., 2013. Ontogenetic functional diversity: size structure of a keystone predator drives functioning of a complex ecosystem. Ecology 94, 1046–1056. https://doi.org/10.1890/12-0378.1

Ruxton, G.D., 2005. Increasing search rate over time may cause a slower than expected increase in prey encounter rate with increasing prey density. Biology Letters 1, 133–135. https://doi.org/10.1098/rsbl.2004.0292

Ruxton, G.D., Allen, W.L., Sherratt, T.N., Speed, M.P., 2004. Avoiding Attack: The Evolutionary Ecology of Crypsis, Aposematism, and Mimicry, Avoiding Attack. Oxford University Press.

Saha, N., Aditya, G., Banerjee, S., Saha, G.K., 2012. Predation potential of odonates on mosquito larvae: implications for biological control. Biological Control 63, 1–8. https://doi.org/10.1016/j.biocontrol.2012.05.004

Samu, F., 1993. Wolf spider feeding strategies: optimality of prey consumption in *Pardosa hortensis*. Oecologia 94, 139–145.

Sarnelle, O., Wilson, A.E., 2008. Type III functional response in *Daphnia*. Ecology 89, 1723–1732. https://doi.org/10.1890/07-0935.1

Schenk, D., Bacher, S., 2002. Functional response of a generalist insect predator to one of its prey species in the field. Journal of Animal Ecology 71, 524–531. https://doi.org/10.1046/j.1365-2656.2002.00620.x

Schmidt, J.M., Sebastian, P., Wilder, S.M., Rypstra, A.L., 2012. The nutritional content of prey affects the foraging of a generalist arthropod predator. PLoS ONE 7, e49223. https://doi.org/10.1371/journal.pone.0049223

Schneider, F.D., Scheu, S., Brose, U., 2012. Body mass constraints on feeding rates determine the consequences of predator loss. Ecology Letters 15, 436–443. https://doi.org/10.1111/j.1461-0248.2012.01750.x

Schoener, T.W., 2011. The newest synthesis: understanding the interplay of evolutionary and ecological dynamics. Science 331, 426–429. https://doi.org/10.1126/science.1193954

Schröder, A., Kalinkat, G., Arlinghaus, R., 2016. Individual variation in functional response parameters is explained by body size but not by behavioural types in a poeciliid fish. Oecologia 182, 1129–1140. https://doi.org/10.1007/s00442-016-3701-7

Secor, S.M., 2008. Digestive physiology of the Burmese python: broad regulation of integrated performance. Journal of Experimental Biology 211, 3767–3774. https://doi.org/10.1242/jeb.023754

Seiedy, M., Saboori, A., Allahyari, H., Talaei-Hassanloui, R., Tork, M., 2012. Functional Response of *Phytoseiulus persimilis* (Acari: Phytoseiidae) on Untreated and *Beauveria bassiana* - Treated Adults of *Tetranychus urticae* (Acari: Tetranychidae). J Insect Behav 25, 543–553. https://doi.org/10.1007/s10905-012-9322-z

Sentis, A., Hemptinne, J.-L., Brodeur, J., 2012. Using functional response modeling to investigate the effect of temperature on predator feeding rate and energetic efficiency. Oecologia 169, 1117–1125. https://doi.org/10.1007/s00442-012-2255-6

Siddiqui, A., Omkar, D., Paul, S.C., Mishra, G., 2015. Predatory responses of selected lines of developmental variants of ladybird, *Propylea dissecta* (Coleoptera: Coccinellidae) in relation to increasing prey and predator densities. Biocontrol Science and Technology 25, 992–1010. https://doi.org/10.1080/09583157.2015.1024101

Sih, A., Christensen, B., 2001. Optimal diet theory: when does it work, and when and why does it fail? Animal Behaviour 61, 379–390. https://doi.org/10.1006/anbe.2000.1592

Sih, A., Englund, G., Wooster, D., 1998. Emergent impacts of multiple predators on prey. Trends in Ecology & Evolution 13, 350–355. https://doi.org/10.1016/S0169-5347(98)01437-2

Skalski, G.T., Gilliam, J.F., 2001. Functional responses with predator interference: viable alternatives to the Holling type II model. Ecology 82, 3083–3092.

Slough, B.G., Mowat, G., 1996. Lynx population dynamics in an untrapped refugium. The Journal of Wildlife Management 60, 946–961. https://doi.org/10.2307/3802397

Smith, N.G., 1969. Provoked release of mobbing-a hunting technique of *Micrastur* falcons. Ibis 111, 241–243. https://doi.org/10.1111/j.1474-919X.1969.tb02530.x

Smout, S., Lindstrøm, U., 2007. Multispecies functional response of the minke whale *Balaenoptera acutorostrata* based on small-scale foraging studies. Marine Ecology Progress Series 341, 277–291. https://doi.org/10.3354/meps341277

Smout, S., Asseburg, C., Matthiopoulos, J., Fernández, C., Redpath, S., Thirgood, S., Harwood, J., 2010. The functional response of a generalist predator. PLoS ONE 5, e10761. https://doi.org/10.1371/journal.pone.0010761

Smout, S., Rindorf, A., Wanless, S., Daunt, F., Harris, M.P., Matthiopoulos, J., 2013. Seabirds maintain offspring provisioning rate despite fluctuations in prey abundance: a multispecies functional response for guillemots in the North Sea. Journal of Applied Ecology 50, 1071–1079. https://doi.org/10.1111/1365-2664.12095

Solomon, M.E., 1949. The natural control of animal populations. Journal of Animal Ecology 18, 1–35. https://doi.org/10.2307/1578

Song, Y.H., Heong, K.L., 1997. Changes in searching responses with temperature of *Cyrtorhinus lividipennis* reuter (Hemiptera: Miridae) on the eggs of the brown planthopper, *Nilaparvata lugens* (Stål.) (Homoptera: Delphacidae). Population Ecology 39, 201–206. https://doi.org/10.1007/BF02765266

Spitze, K., 1985. Functional response of an ambush predator: *Chaoborus americanus* predation on *Daphnia pulex*. Ecology 66, 938–949. https://doi.org/10.2307/1940556

Stafstrom, J.A., Hebets, E.A., 2016. Nocturnal foraging enhanced by enlarged secondary eyes in a net-casting spider. Biology Letters 12, 20160152. https://doi.org/10.1098/rsbl.2016.0152

Stahlecker, D.W., Mikesic, D.G., White, J.N., Shaffer, S., DeLong, J.P., Blakemore, M.R., Blakemore, C.E., 2009. Prey remains in nests of four corners golden eagles, 1998–2008. Western Birds 40, 301–306.

Stearns, S.C., 1992. The Evolution of Life Histories. Oxford University Press.

Strauss, S.Y., Irwin, R.E., 2004. Ecological and evolutionary consequences of multispecies plant-animal interactions. Annual Review of Ecology, Evolution, and Systematics 35, 435–466. https://doi.org/10.1146/annurev.ecolsys.35.112202.130215

Streams, F.A., 1994. Effect of prey size on attack components of the functional response by *Notonecta undulata*. Oecologia 98, 57–63. https://doi.org/10.1007/BF00326090

Tallian, A., Smith, D.W., Stahler, D.R., Metz, M.C., Wallen, R.L., Geremia, C., Ruprecht, J., Wyman, C.T., MacNulty, D.R., 2017. Predator foraging response to a resurgent dangerous prey. Functional Ecology 31, 1418–1429. https://doi.org/10.1111/1365-2435.12866

Tenhumberg, B., 1995. Estimating predatory efficiency of *Episyrphus balteatus* (Diptera: Syrphidae) in cereal fields. Environmental Entomology 24, 687–691. https://doi.org/10.1093/ee/24.3.687

Thieltges, D.W., Amundsen, P.-A., Hechinger, R.F., Johnson, P.T.J., Lafferty, K.D., Mouritsen, K.N., Preston, D.L., Reise, K., Zander, C.D., Poulin, R., 2013. Parasites as prey in aquatic food webs: implications for predator infection and parasite transmission. Oikos 122, 1473–1482. https://doi.org/10.1111/j.1600-0706.2013.00243.x

Thompson, D.J., 1975. Towards a predator-prey model incorporating age structure: The effects of predator and prey size on the predation of *Daphnia magna* by *Ischnura elegans*. Journal of Animal Ecology 44, 907–916. https://doi.org/10.2307/3727

Thompson, D.J., 1978. Towards a realistic predator-prey model: the effect of temperature on the functional response and life history of larvae of the damselfly, *Ischnura elegans*. Journal of Animal Ecology 47, 757–767. https://doi.org/10.2307/3669

Thorp, C.J., Alexander, M.E., Vonesh, J.R., Measey, J., 2018. Size-dependent functional response of *Xenopus laevis* feeding on mosquito larvae. PeerJ 6, e5813. https://doi.org/10.7717/peerj.5813

Tollrian, R., 1995. Predator-induced morphological defenses: costs, life history shifts, and maternal effects in *Daphnia pulex*. Ecology 76, 1691–1705. https://doi.org/10.2307/1940703

Toscano, B.J., Griffen, B.D., 2014. Trait-mediated functional responses: predator behavioural type mediates prey consumption. Journal of Animal Ecology 83, 1469–1477. https://doi.org/10.1111/1365-2656.12236

Toscano, B.J., Newsome, B., Griffen, B.D., 2014. Parasite modification of predator functional response. Oecologia 175, 345–352. https://doi.org/10.1007/s00442-014-2905-y

Trexler, J.C., McCulloch, C.E., Travis, J., 1988. How can the functional reponse best be determined? Oecologia 76, 206–214.

Tully, T., Cassey, P., Ferrière, R., 2005. Functional response: rigorous estimation and sensitivity to genetic variation in prey. Oikos 111, 479–487. https://doi.org/10.1111/j.1600-0706.2005.14062.x

Turchin, P., 2001. Does population ecology have general laws? Oikos 94, 17–26.

Tyutyunov, Y., Titova, L., Arditi, R., 2008. Predator interference emerging from trophotaxis in predator–prey systems: an individual-based approach. Ecological Complexity 5, 48–58. https://doi.org/10.1016/j.ecocom.2007.09.001

Uiterwaal, S.F., DeLong, J.P., 2018. Multiple factors, including arena size, shape the functional responses of ladybird beetles. Journal of Applied Ecology 55, 2429–2438. https://doi.org/10.1111/1365-2664.13159

Uiterwaal, S.F., DeLong, J.P., 2020. Functional responses are maximized at intermediate temperatures. Ecology 101, e02975. https://doi.org/10.1002/ecy.2975

Uiterwaal, S.F., Mares, C., DeLong, J.P., 2017. Body size, body size ratio, and prey type influence the functional response of damselfly nymphs. Oecologia 185, 339–346. https://doi.org/10.1007/s00442-017-3963-8

Uiterwaal, S.F., Lagerstrom, I.T., Lyon, S.R., DeLong, J.P., 2018. Data paper: FoRAGE (Functional Responses from Around the Globe in all Ecosystems) database: a compilation of functional responses for consumers and parasitoids. bioRxiv 503334. https://doi.org/10.1101/503334

Uiterwaal, S.F., Dell, A.I., DeLong, J.P., 2019. Arena size modulates functional responses via behavioral mechanisms. Behavioral Ecology 30, 483–489. https://doi.org/10.1093/beheco/ary188

Urban, M.C., Freidenfelds, N.A., Richardson, J.L., 2020. Microgeographic divergence of functional responses among salamanders under antagonistic selection from apex predators. Proceedings of the Royal Society B: Biological Sciences 287, 20201665. https://doi.org/10.1098/rspb.2020.1665

Uszko, W., Diehl, S., Pitsch, N., Lengfellner, K., Müller, T., 2015. When is a type III functional response stabilizing? Theory and practice of predicting plankton dynamics under enrichment. Ecology 96, 3243–3256. https://doi.org/10.1890/15-0055.1

Uszko, W., Diehl, S., Englund, G., Amarasekare, P., 2017. Effects of warming on predator–prey interactions—a resource-based approach and a theoretical synthesis. Ecology Letters 20, 513–523. https://doi.org/10.1111/ele.12755

Uszko, W., Diehl, S., Wickman, J., 2020. Fitting functional response surfaces to data: a best practice guide. Ecosphere 11, e03051. https://doi.org/10.1002/ecs2.3051

Vallina, S.M., Ward, B.A., Dutkiewicz, S., Follows, M.J., 2014. Maximal feeding with active prey-switching: a kill-the-winner functional response and its effect on global diversity and biogeography. Progress in Oceanography 120, 93–109. https://doi.org/10.1016/j.pocean.2013.08.001

van den Bosch, F., Santer, B., 1993. Cannibalism in *Cyclops abyssorum*. Oikos 67, 19–28. https://doi.org/10.2307/3545091

Van Valen, L., 1973. A new evolutionary law. Evolutionary Theory 1, 1–33.

Vermont, A.I., Martínez, J.M., Waller, J.D., Gilg, I.C., Leavitt, A.H., Floge, S.A., Archer, S.D., Wilson, W.H., Fields, D.M., 2016. Virus infection of *Emiliania huxleyi* deters grazing by the copepod *Acartia tonsa*. Journal of Plankton Research 38, 1194–1205. https://doi.org/10.1093/plankt/fbw064

Vucic-Pestic, O., Rall, B.C., Kalinkat, G., Brose, U., 2010. Allometric functional response model: body masses constrain interaction strengths. Journal of Animal Ecology 79, 249–256. https://doi.org/10.1111/j.1365-2656.2009.01622.x

Wahlström, E., Persson, L., Diehl, S., Byström, P., 2000. Size-dependent foraging efficiency, cannibalism and zooplankton community structure. Oecologia 123, 138–148.

Walker, S.E., Rypstra, A.L., 2002. Sexual dimorphism in trophic morphology and feeding behavior of wolf spiders (Araneae: Lycosidae) as a result of differences in reproductive roles. The Canadian Journal of Zoology 80, 679–688. https://doi.org/10.1139/z02-037

Ware, D.M., 1972. Predation by rainbow trout (*Salmo gairdneri*): the influence of hunger, prey density, and prey size. Journal of the Fisheries Board of Canada 29, 1193–1201. https://doi.org/10.1139/f72-175

Wasserman, R.J., Alexander, M.E., Weyl, O.L.F., Barrios-O'Neill, D., Froneman, P.W., Dalu, T., 2016. Emergent effects of structural complexity and temperature on predator–prey interactions. Ecosphere 7, e01239. https://doi.org/10.1002/ecs2.1239

Webster, J.P., 2007. The effect of *Toxoplasma gondii* on animal behavior: playing cat and mouse. Schizophrenia Bulletin 33, 752–756. https://doi.org/10.1093/schbul/sbl073

Weitz, J.S., Levin, S.A., 2006. Size and scaling of predator-prey dynamics. Ecology Letters 9, 548–557. https://doi.org/10.1111/j.1461-0248.2006.00900.x

Welsh, J.E., Steenhuis, P., Moraes, K.R. de, Meer, J. van der, Thieltges, D.W., Brussaard, C.P.D., 2020. Marine virus predation by non-host organisms. Scientific Reports 10, 1–9. https://doi.org/10.1038/s41598-020-61691-y

Weterings, R., Umponstira, C., Buckley, H.L., 2015. Density-dependent allometric functional response models. Ecological Modelling 303, 12–18. https://doi.org/10.1016/j.ecolmodel.2015.02.003

Whittington, J., St. Clair, C.C., Mercer, G., 2005. Spatial responses of wolves to roads and trails in mountain valleys. Ecological Applications 15, 543–553.

Wilder, S.M., Rypstra, A.L., 2008. Diet quality affects mating behaviour and egg production in a wolf spider. Animal Behaviour 76, 439–445. https://doi.org/10.1016/j.anbehav.2008.01.023

Wilder, S.M., Le Couteur, D.G., Simpson, S.J., 2013. Diet mediates the relationship between longevity and reproduction in mammals. AGE 35, 921–927. https://doi.org/10.1007/s11357-011-9380-8

Wilhelm, F.M., Schindler, D.W., McNaught, A.S., 2000. The influence of experimental scale on estimating the predation rate of *Gammarus lacustris* (Crustacea: Amphipoda) on *Daphnia* in an alpine lake. Journal of Plankton Research 22, 1719–1734. https://doi.org/10.1093/plankt/22.9.1719

Williams, F.M., Juliano, S.A., 1996. Functional responses revisited. Environmental Entomology 25, 549–550. https://doi.org/10.1093/ee/25.3.549

Williams, R.J., Martinez, N.D., 2004. Stabilization of chaotic and non-permanent food-web dynamics. European Physical Journal B 38, 297–303. https://doi.org/10.1140/epjb/e2004-00122-1

Wilmers, C.C., Estes, J.A., Edwards, M., Laidre, K.L., Konar, B., 2012. Do trophic cascades affect the storage and flux of atmospheric carbon? An analysis of sea otters and kelp forests. Frontiers in Ecology and the Environment 10, 409–415. https://doi.org/10.1890/110176

Wong, M.C., Barbeau, M.A., Dowd, M., Richard, K.R., 2006. Behavioural mechanisms underlying functional response of sea stars *Asterias vulgaris* preying on juvenile sea scallops *Placopecten magellanicus*. Marine Ecology Progress Series 317, 75–86. https://doi.org/10.3354/meps317075

Wootton, J.T., Emmerson, M., 2005. Measurement of interaction strength in nature. Annual Review of Ecology, Evolution, and Systematics 36, 419–444. https://doi.org/10.1146/annurev.ecolsys.36.091704.175535

Yaşar, B., Özger, Ş., 2005. Development, feeding and reproduction responses of *Adalia fasciatopunctata revelierei* (Mulsant) (Coleoptera: Coccinellidae) to *Hyalopterus pruni* (Geoffroy) (Homoptera: Aphididae). Journal of Pest Science 78, 199–203. https://doi.org/10.1007/s10340-005-0089-2

Yoshida, T., Jones, L.E., Ellner, S.P., Fussmann, G.F., Hairston, Jr., N.G., 2003. Rapid evolution drives ecological dynamics in a predator–prey system. Nature 424, 303–306. https://doi.org/10.1038/nature01767

Index